Programming the Basic Atom Microcontroller

By Chuck Hellebuyck

The publisher offers special discounts on bulk orders of this book.
For information contact:

Electronic Products
P.O. Box 251
Milford, MI 48381
www.elproducts.com
chuck@elproducts.com

The Microchip name and logo, MPLAB® and PIC® are registered trademarks of Microchip Technology Inc. in the U.S.A. and other countries.
BasicAtom is a registered trademark of Basic Micro Inc.

All other trademarks mentioned herein are property of their respective companies.

Printed in United States of America
Cover design by Rich Sherlitz (rich@electricchili.com)

Dedication
This book is dedicated to my wife Erin and my children Chris, Connor and Brittany.

Thanks for letting dad spend all those hours in front of my computer.

I also dedicate this book to my father who supported me in my early years of life when I caught the electronics bug. My father also taught me the electrical basics that launched my career. I just wish I could have used it to design a cure for his cancer.

Table of Contents

Introduction

I've been programming Microchip PIC microcontrollers for years but the day I saw the Atom microcontroller and Atom modules from BasicMicro, I knew I had found the greatest platform for the beginner, hobbyist and even professional electronic developer. You cannot read an electronics hobbyist magazine without seeing a project that is microcontroller based. The evolution of the homemade robot has advanced due to the advancements and affordability of microcontrollers. But for some it's still difficult to get started or to find all the right pieces you need to build your own microcontroller development lab.

The Atom micro and Atom modules fill that desire very nicely and give you all the power of a high priced software compiler and development platform in a single, low cost package. The best part is you have numerous options to choose from. From individual chips to small DIP modules to complex full featured development boards, all of them can be programmed with the Atom BASIC language compiler that you can download for free from the BasicMicro website.

This doesn't mean you won't someday advance your skills to needing professional development tools for high volume production such as a multiple PIC version BASIC compiler or assembly, but why overwhelm yourself and your wallet when most users only need what the Atom can deliver.

I've sold many development platforms at a discount from my website with the single purpose of helping others get started programming Microchip PIC microcontrollers. I do this because I love this stuff and want to share it with others. There are many hardware developers that sat out the microcontroller craze because it was such a big step up from the TTL chips most had been using in their electronic projects. That big step was made smaller because of the Atom.

My first book "Programming PIC Microcontrollers With PicBasic" was written to help individuals learn how to program Microchip PIC microcontrollers with the PICBasic compilers from microEngineering Labs. It was received so well I decided to introduce others to the Atom modules through a second book and this is it. Atoms do not replace PICBasic and PICBasic does not replace the Atom. They are just different levels of stepping-stones that allow users from beginner to advanced deliver custom microcontroller based designs easily and at low cost. In fact, BasicMicro also offers the MBasic Basic Compiler that competes directly with the PICBasic compilers. The Atom and MBasic share the same format and 99% of the same commands. This means learning the Atom could be the catalyst to get you started into this wonderful world of electronic design. You never know, the next PC revolution could be started in your basement lab and if this book or my other writings helped get you started then every finger numbing word I typed was worth it.

If you have any questions regarding this book or the projects in this book, you can usually get me via email at chuck@elproducts.com. My website is also dedicated to programming Microchip PICs in Basic so you will probably find various helpful tips there as well. I hope this book is everything you hoped it would be and more. Now lets get started learning how to use these Atom modules.

Chapter 1 – Basic Atom Microcontroller

What is an Atom Module

You probably already know what an Atom module looks like and what it basically does otherwise you wouldn't have bought this book. To completely understand all the capabilities of the Atom though, it helps to dive even deeper than what you see and get into the guts of an Atom module and see what makes it tick. This chapter intends to explain all the technical details of the Atom in simple terminology so both beginners and experienced users can benefit.

The Atom module is really a miniature computer control system. It contains a Microchip PIC Microcontroller with the Atom self-programming bootloader software burned in, also known as an Atom interpreter chip or as I often refer to it in this book as the Atom micro. The module also includes a 5-volt regulator circuit, a ceramic resonator clock oscillator, communication circuitry used to download a BASIC program into the Atom's microcontroller memory, and the Input/Output (I/O) interface.

Atom Module Architecture

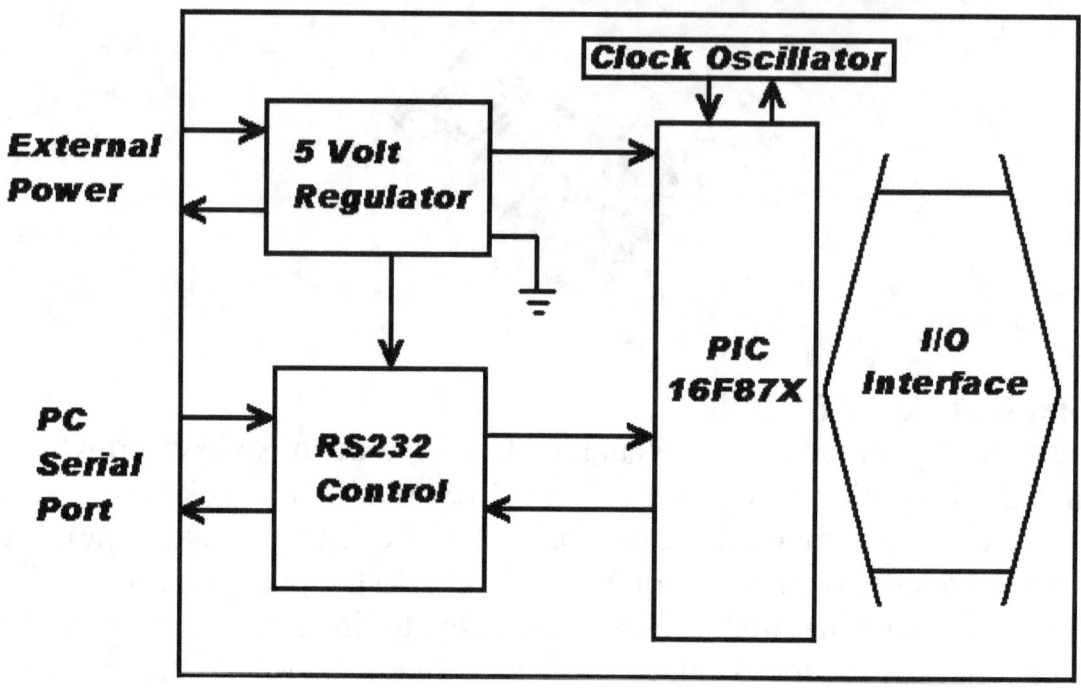

There are three versions of the Atom micro, a 40-pin and two 28-pin versions (an 18 pin version is also in development). The 40-pin Atom micro is used in the 40-pin Atom module and BasicBoard development board. The 40-pin Atom micro is built from a 16F877A or 16F887 version of the Microchip PIC. The Atom OEM, and 24 pin modules use the 28 pin version-A Atom firmware chip based on the Microchip PIC 16F876A or 16F886. The Ultimate OEM and 28-pin modules use the 28-pin version-B Atom firmware chip also based on the PIC 16F876A. The 877A and 876A are identical inside except the 40 pin 16F877 has more I/O pins. The 16F887 and 16F886 are updated versions of the 16F877A and 16F876A so some of the newer Atom chips will be the 16F88x family. The new 18 pin being developed is based on the 16F88 18 pin part. This has not been released yet as this book is being published but should be released soon.

For the rest of this book though I'll focus on the 16F87xA parts as these are most the most common.

Basic Atom 40 pin Interpreter Chip

The Atom micro takes advantage of all the Microchip PIC microcontroller features and offers the following:

- **8K of Program Space**
- **368 Bytes of User / System RAM**
- **256 Bytes of User EEPROM**
- **33,000 Plus Instruction Per Second**
- **Three Hardware Timers**
- **Two Capture, Compare peripherals**
- **Two PWM modules (10-bit)**
- **10-bit Analog-to-Digital converters**
- **Built in Hardware Serial Port**
- **Interrupt Capable**
- **32 x 32 Bit Math**
- **Floating Point Math**
- **Variables with values up to 4,294, 967,295**

The 8k of program space is flash program memory, which can be programmed over and over more than 10,000 times. Inside the Atom micro is all the circuitry that makes the Atom module I/O work. It has the input/output circuitry that can drive Light Emitting Diodes (LED), small relays, transistor circuits, robotic sensors, Liquid Crystal Displays (LCDs), and numerous other electronic devices. So to really understand the Atom micro it's best to understand the PIC Microcontroller.

15

What is a Microchip PIC?

Microchip is a company that created an 8-bit microcontroller called the PIC that could be programmed to perform an infinite amount of functions. To best understand what that means let's first explain what a microcontroller is.

Everybody reading this has probably used a computer run by a microprocessor. The PCs central microprocessor has several support items that allow it to function. The memory, where programs are stored, known as a hard drive or ROM. The RAM, or temporary memory used by the programs running in the microprocessor. And finally, the interface to the outside world through input and output control, also known as the BIOS or I/O.

Through the I/O the PC sends information to be displayed on the screen you watch, or the printer you send documents to. The I/O also reads the keyboard and mouse position. Basically everything the PC does with a useful purpose to human's runs through the I/O. That describes what a PC is but, what if you could shrink all those components, microprocessor, ROM, RAM and I/O, into a single integrated circuit?

Well, it can be done and that is what a microcontroller is, a miniature computer in a single integrated circuit with a small amount of ROM and RAM and lots of I/O. Microchip PICs are one of the best versions of microcontroller's and the Atom micro's are built from them.

How Does a Microcontroller Work?

A microcontroller requires a series of coded electrical signals to be stored into its ROM, which controls the micro's I/O. These electrical signals are also known as the 1's and 0's of binary code. This is also known as software or program code. When a microcontroller is said to be programmed or have code burned into it, it is getting these coded electrical signals stored into its ROM.

To function a code, the microcontroller needs a way to select each command from ROM one at a time, which is known as running a program. To do this, the microcontroller requires a clock oscillator, also known as a crystal or resonator clock that sends a continuous pulse to the microcontroller's central circuitry. This is very similar to a PC that advertises a 1 Gigahertz Pentium processor. The speed of the PIC in the Atom is only 20 Megahertz but plenty fast enough for most projects.

When the micro is first powered up the resonator clock starts pulsing the same way our heart pulses our blood through our body. On each pulse of the clock, the micro retrieves a new command from ROM to execute on the I/O. By arranging these binary codes properly you can make the I/O turn pins on and off to control electrical circuitry connected to the I/O pins. That circuitry could be a simple relay that turns a light on during the night and off during the day or it could be more complex and control the motors of a robot while reading an obstacle sensor. All you need to do is write this series of binary code properly, which is the software. To make it easier to develop this binary code, the Atom micro comes with a BASIC language compiler.

What is a Basic Language Compiler?

A compiler is a PC software application that converts easy to read and understand words (BASIC commands) into an assembly language file and then assembles that into the binary code the micro needs. Binary code is the lowest level of software and BASIC is considered a high level language. Once that binary code file is created, then the Microchip PIC in the Atom can be programmed.

Basic language compilers typically cost $100 to $300 and can program various Microchip PICs but the Atom Basic Compiler is given away for free by Basic Micro. They can do this because they limited it to only work with the Atom micro. They lose money on the compiler but make it up on the sale of Atom interpreter chips and modules. Therefore the Atom user gets a full-featured Basic Compiler for the price of an Atom module or interpreter chip that starts at $20. The move to the 16F88x family of parts allows Basic Micro to offer a cheaper price than $20 on the interpreter chips. They are also looking to give the interpreter its own name and call it the Atom Nano. This book was published before this was finalized so watch for that lower cost Atom Nano chip. Even at $20 per chip though it's a lot cheaper than spending $300 to get started programming Microchip PIC using one of the other options.

How does the PIC actually get programmed in the Atom?

There are really two ways to load a binary program into a Microchip PIC, 1) with a PIC hardware programmer or 2) let the PIC program itself using a bootloader (Atom module method).

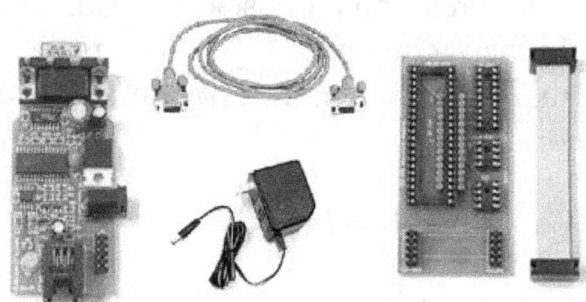

Option 1
A hardware programmer is a custom designed module that receives the binary code file created by the compiler via a PC serial or parallel port. The PIC Programmer generates the electrical signals the Microchip PIC needs to see and then the binary file is downloaded to the PIC program memory. Below is an ISP-Pro Hardware PIC Programmer from BasicMicro that performs this function. The socketed adapter is attached to this programmer for the PIC to be plugged into while the power adapter and serial cable are used to supply the power and the connection to the PC.

Option 2 (Atom Module Method)

A bootloader is a custom binary file created in assembly language that allows a PIC to program itself without use of a hardware PIC programmer like the one shown in option 1. To get it to work though the bootloader binary file must first be programmed into the PIC using a hardware PIC programmer such as the ISP-Pro. So in other words you need option 1 in order to get to option 2. But the Atom chip has that already programmed in thus eliminating the need for you to buy the PIC programming hardware of Option 1.

Basic Micro created a custom bootloader for the Microchip PIC and pre-programs it into every Atom interpreter chip before they are built into the Atom module. The Atom bootloader is designed to only receive Atom BASIC Compiler files and it receives them through a PC serial port connection.

When the Atom module is first powered up it runs the bootloader software that watches for new software to be downloaded from the PC serial port. It watches for roughly ½ second and if no software is ready to download, then it switches over to run mode and starts executing whatever program was previously loaded.

This should help you understand the Atom Interpreter or Atom micro, as I like to call it. This should also help you understand that you don't need a full module to run the Atom; all you really need is the Atom interpreter chip and some support circuitry. This is a great option because once you have a design worked out you can then produce several prototypes for a cheaper cost than if you had to buy a bunch of modules. But many people like to have all the support circuitry built into one unit and the modules do just that.

BasicATOM24-M   BasicATOM

BasicATOM28-M   BasicATOM40-M

Atom Modules

Basic Micro Atom modules come in four versions, 24-pin, 28-pin, 40-pin and OEM. The 24, 28 and 40 pin modules are DIP packages that fit in a 0.600" wide socket. These are all built using surface mount device (SMD) versions of the Microchip PICs. This allows them to be built small. This also makes them more difficult to repair if you happen to fry the on-board regulator (which can happen).

The OEM version is a leaded version with I/O similar to the 24-pin version but adds a few solder pads to access the A/D pins. Because it is leaded, the Atom micro interpreter chip can be removed and built into a permanent design and replaced with a new Atom interpreter chip. This makes it sort of an Atom micro programmer. Because it's limited to the 24 pin I/O the extra I/O, which are A/D pins, have to be wired and jumpered over if you use the module in a breadboard. I took the OEM concept and developed my own version the "Ultimate OEM module" shown below.

Ultimate OEM

All the A/D ports are brought out to the 25-pin header so you have easy access to these extra pins. I also added a serial LCD header, momentary switch tied to P0, LEDs tied to P1 and P2. I also added a power adapter connection port with on/off switch and power on indicator LED. I use it all the time for my experiments and I use it in all the projects in this book.

Finally I wanted to mention my BasicBoard development module. For those that are not interested in building all the hardware connections, the

BasicBoard uses the 40 pin Atom micro and pre-connects several features including an LCD, 4 momentary switches, 8 LEDs, speaker, potentiometer, and 3 pin headers for connecting sensors or servo motors or anything else you want to build. This has been a popular unit for beginners and colleges teaching programming.

Each of these modules uses the same Atom Basic compiler software. This makes it easy to move from design to design without tearing up your code. The next several pages contain charts that describe the I/O and pin-out details for all the Atom modules. This should be a good reference for you. Each chart contains pin numbers for the modules and the coordinating pins on the Atom interpreter PIC Chip. This should help you understand how to apply Microchip PIC information to the Atom module.

Atom 24-pin Module Pin-out Description

Pin Name	Atom 24-pin Module Pin Number	PIC 16F876 Pin Name	PIC 16F876, Atom28A-IC Pin Number	Function
P0	5	B0	21	Digital I/O and External Interrupt Pin
P1	6	B1	22	Digital I/O
P2	7	B2	23	Digital I/O
P3	8	B3	24	Digital I/O
P4	9	B4	25	Digital I/O
P5	10	B5	26	Digital I/O
P6	11	B6	27	Digital I/O – PIC Programming Pin
P7	12	B7	28	Digital I/O – PIC Programming Pin
P8	13	C0	11	Digital I/O or Timer1 Clock Input / Timer1 Oscillator Output
P9	14	C1	12	Digital I/O or Capture2 input/ Compare2 output HPWM2 output Timer Oscillator Input
P10	15	C2	13	Digital I/O or Capture1 input Compare1 output HPWM1 output

P11	16	C3	14	Digital I/O or Serial SPI/I2C Clock
P12	17	C4	15	Digital I/O or Serial SPI /I2C Data In
P13	18	C5	16	Digital I/O or Synchronous Serial Port Output
P14	19	C6	17	Digital I/O or Asynchronous Transmit or Synchronous Clock
P15	20	C7	18	Digital I/O or Asynchronous Receive or Synchronous Data
AX0	Pad	A2	4	Digital I/O or Analog to Digital input Also Negative voltage reference for A/D port.
AX1	Pad	A3	5	Digital I/O or Analog to Digital input Also Positive voltage reference for A/D port.
AX2	Pad	A4	6	Open Drain Digital I/O or Timer0 Input
AX3	Pad	A5	7	Digital I/O or Analog to Digital input

ATN	3	MCLR	1	Master Reset Pin or Programming Voltage Input
Vdd	21	Vdd	20	Positive Supply for Logic and I/O Pins
Vss	4, 23	Vss	8, 19	Ground Reference for Logic and I/O Pins
Vin	24	N/A	N/A	Atom Module Power Input
Res	22	MCLR	1	Master Reset Pin or Programming Voltage Input
S_in	2	RA0	2	Atom Programming Pin
S_out	1	RA1	3	Atom Programming Pin

Atom 28-pin Module Pin-out Description

Pin Name	Atom 28-pin Module Pin Number	PIC 16F876 Pin Name	PIC 16F876, Atom28B-IC Pin Number	Function
P0	5	B0	21	Digital I/O and External Interrupt Pin
P1	6	B1	22	Digital I/O
P2	7	B2	23	Digital I/O
P3	8	B3	24	Digital I/O
P4	9	B4	25	Digital I/O
P5	10	B5	26	Digital I/O
P6	11	B6	27	Digital I/O – PIC Programming Pin
P7	12	B7	28	Digital I/O – PIC Programming Pin
P8	17	C0	11	Digital I/O or Timer1 Clock Input / Timer1 Oscillator Output

P9	18	C1	12	Digital I/O or Capture2 input/ Compare2 output HPWM2 output Timer Oscillator Input
P10	19	C2	13	Digital I/O or Capture1 input Compare1 output HPWM1 output
P11	20	C3	14	Digital I/O or Serial SPI/I2C Clock
P12	21	C4	15	Digital I/O or Serial SPI /I2C Data In
P13	22	C5	16	Digital I/O or Synchronous Serial Port Output
P14	23	C6	17	Digital I/O or Asynchronous Transmit or Synchronous Clock
P15	24	C7	18	Digital I/O or Asynchronous Receive or Synchronous Data
AX0	13	A0	2	Digital I/O or Analog to Digital input
AX1	14	A1	3	Digital I/O or Analog to Digital input

AX2	16	A2	4	Digital I/O or Analog to Digital input Also Negative voltage reference for A/D port.
AX3	15	A3	5	Digital I/O or Analog to Digital input Also Positive voltage reference for A/D port.
ATN	3	MCLR	1	Master Reset Pin or Programming Voltage Input
Vdd	25	Vdd	20	Positive Supply (5v) for Logic and I/O Pins
Vss	4, 27	Vss	8, 19	Ground Reference for Logic and I/O Pins
Vin	28	N/A	N/A	Atom Module Power Input (9-12v) Feeds Vdd thru 5v Regulator
Res	26	MCLR	1	Master Reset Pin or Programming Voltage Input
S_in	2	RA4	6	Atom Programming Pin
S_out	1	RA4	6	Atom Programming Pin

Atom 40-pin Module Pin-out Description

Pin Name	Atom 40-pin Module Pin Number	PIC 16F877 Pin Name	PIC 16F877, Atom40-IC Pin Number	Function
P0	5	B0	33	Digital I/O and External Interrupt Pin
P1	6	B1	34	Digital I/O
P2	7	B2	35	Digital I/O
P3	8	B3	36	Digital I/O
P4	9	B4	37	Digital I/O
P5	10	B5	38	Digital I/O
P6	11	B6	39	Digital I/O – PIC Programming Pin
P7	12	B7	40	Digital I/O – PIC Programming Pin
P8	29	C0	15	Digital I/O or Timer1 Clock Input / Timer1 Oscillator Output
P9	30	C1	16	Digital I/O or Capture2 input/ Compare2 output HPWM2 output Timer Oscillator Input
P10	31	C2	17	Digital I/O or Capture1 input Compare1 output HPWM1 output

P11	32	C3	18	Digital I/O or Serial SPI/I2C Clock
P12	33	C4	23	Digital I/O or Serial SPI /I2C Data In
P13	34	C5	24	Digital I/O or Synchronous Serial Port Output
P14	35	C6	25	Digital I/O or Asynchronous Transmit or Synchronous Clock
P15	36	C7	26	Digital I/O or Asynchronous Receive or Synchronous Data
P16	15	D0	19	Digital I/O
P17	16	D1	20	Digital I/O
P18	17	D2	21	Digital I/O
P19	18	D3	22	Digital I/O
P20	19	D4	27	Digital I/O
P21	20	D5	28	Digital I/O
P22	21	D6	29	Digital I/O
P23	22	D7	30	Digital I/O
P24	23	E0	8	Digital I/O or Analog to Digital input
P25	24	E1	9	Digital I/O or Analog to Digital input
P26	25	E2	10	Digital I/O or Analog to Digital input

P27 / AX4	26	A5	7	Digital I/O or Analog to Digital input
P28 / AX0	13	A0	2	Digital I/O or Analog to Digital input
P29 / AX1	14	A1	3	Digital I/O or Analog to Digital input
P30 / AX2	28	A2	4	Digital I/O or Analog to Digital input Also Negative voltage reference for A/D port.
P31 / AX3	27	A3	5	Digital I/O or Analog to Digital input Also Positive voltage reference for A/D port.
Vdd	37	Vdd	11, 32	Positive Supply (5v) for Logic and I/O Pins
Vss	4, 39	Vss	12, 31	Ground Reference for Logic and I/O Pins
Vin	40	N/A	N/A	Atom Module Power Input (9-12v) Feeds Vdd thru 5v Regulator
Res	38	MCLR	1	Master Reset Pin or Programming Voltage Input
S_in	2	A4	6	Atom Programming Pin

| S_out | 1 | A4 | 6 | Atom Programming Pin |
| ATN | 3 | MCLR | 1 | Master Reset Pin or Programming Voltage Input |

OEM Atom Module Pin-out Description

Pin Name	OEM Atom Module Pin Number	PIC 16F876 Pin Name	PIC 16F876, Atom28A-IC Pin Number	Function
P0	5	B0	21	Digital I/O and External Interrupt Pin
P1	6	B1	22	Digital I/O
P2	7	B2	23	Digital I/O
P3	8	B3	24	Digital I/O
P4	9	B4	25	Digital I/O
P5	10	B5	26	Digital I/O
P6	11	B6	27	Digital I/O – PIC Programming Pin
P7	12	B7	28	Digital I/O – PIC Programming Pin

P8	13	C0	11	Digital I/O or Timer1 Clock Input / Timer1 Oscillator Output
P9	14	C1	12	Digital I/O or Capture2 input/ Compare2 output HPWM2 output Timer Oscillator Input
P10	15	C2	13	Digital I/O or Capture1 input Compare1 output HPWM1 output
P11	16	C3	14	Digital I/O or Serial SPI/I2C Clock
P12	17	C4	15	Digital I/O or Serial SPI /I2C Data In
P13	18	C5	16	Digital I/O or Synchronous Serial Port Output
P14	19	C6	17	Digital I/O or Asynchronous Transmit or Synchronous Clock
P15	20	C7	18	Digital I/O or Asynchronous Receive or Synchronous Data

AX0	Solder Pad	A2	4	Digital I/O or Analog to Digital input Also Negative voltage reference for A/D port.
AX1	Solder Pad	A3	5	Digital I/O or Analog to Digital input Also Positive voltage reference for A/D port.
AX2	Solder Pad	A4	6	Open Drain Digital I/O or Timer0 Input
AX3	Solder Pad	A5	7	Digital I/O or Analog to Digital input
ATN	DB9 Connector Pin 4	MCLR	1	Master Reset Pin or Programming Voltage Input
Vdd	3	Vdd	20	Positive Supply for Logic and I/O Pins
Vss	2	Vss	8, 19	Ground Reference for Logic and I/O Pins
Vin	1	N/A	N/A	Atom Module Power Input
Res	4	MCLR	1	Master Reset Pin or Programming Voltage Input
S_in	DB9 Connector Pin 3	RA0	2	Atom Programming Pin
S_out	DB9 Connector Pin 2	RA1	3	Atom Programming Pin

Ultimate OEM Atom Module Pin-out Description

Pin Name	Ultimate OEM Module Pin Number	PIC 16F876 Pin Name	PIC 16F876, Atom28B-IC Pin Number	Function
P0	5	B0	21	Digital I/O and External Interrupt Pin
P1	6	B1	22	Digital I/O
P2	7	B2	23	Digital I/O
P3	8	B3	24	Digital I/O
P4	9	B4	25	Digital I/O
P5	10	B5	26	Digital I/O
P6	11	B6	27	Digital I/O – PIC Programming Pin
P7	12	B7	28	Digital I/O – PIC Programming Pin
P8	13	C0	11	Digital I/O or Timer1 Clock Input / Timer1 Oscillator Output

P9	14	C1	12	Digital I/O or Capture2 input/ Compare2 output HPWM2 output Timer Oscillator Input
P10	15	C2	13	Digital I/O or Capture1 input Compare1 output HPWM1 output
P11	16	C3	14	Digital I/O or Serial SPI/I2C Clock
P12	17	C4	15	Digital I/O or Serial SPI /I2C Data In
P13	18	C5	16	Digital I/O or Synchronous Serial Port Output
P14	19	C6	17	Digital I/O or Asynchronous Transmit or Synchronous Clock
P15	20	C7	18	Digital I/O or Asynchronous Receive or Synchronous Data
AX0	21	A0	2	Digital I/O or Analog to Digital input
AX1	22	A1	3	Digital I/O or Analog to Digital input

AX2	23	A2	4	Digital I/O or Analog to Digital input Also Negative voltage reference for A/D port.
AX3	24	A3	5	Digital I/O or Analog to Digital input Also Positive voltage reference for A/D port.
A4	25	A4	6	Programming Pin for Atom Timer0 input for counter Also Open Drain Output **Note: Use of this pin can prevent in-circuit programming**
A5	On 6-pin Header	A5	7	Not used with Atom Chip
ATN	DB9 Connector Pin 4	MCLR	1	Master Reset Pin or Programming Voltage Input
Vdd	3	Vdd	20	Positive Supply for Logic and I/O Pins
Vss	2	Vss	8, 19	Ground Reference for Logic and I/O Pins
Vin	1	N/A	N/A	Atom Module Power Input
Res	4	MCLR	1	Master Reset Pin or Programming Voltage Input

| *S_in* | DB9 Connector Pin 3 | RA4 | 6 | Atom Programming Pin |
| *S_out* | DB9 Connector Pin 2 | RA4 | 6 | Atom Programming Pin |

BasicBoard Pin-out Description

Pin Name	BasicBoard Module Pin Number	PIC 16F877 Pin Name	PIC 16F877, Atom40-IC Pin Number	Function
P0	5	B0	33	LED0
P1	6	B1	34	LED1
P2	7	B2	35	LED2
P3	8	B3	36	LED3
P4	9	B4	37	LED4
P5	10	B5	38	LED5
P6	11	B6	39	LED6
P7	12	B7	40	LED7
P8	29	C0	15	LCD DB4 pin
P9	30	C1	16	LCD DB5 pin
P10	31	C2	17	LCD DB6 pin
P11	32	C3	18	LCD DB7 pin
P12	33	C4	23	SW3 (switch 3)
P13	34	C5	24	SW4 (switch 4)
P14	35	C6	25	External RS232 Tx pin
P15	36	C7	26	External RS232 Rx pin
P16	15	D0	19	LCD E pin
P17	16	D1	20	LCD RS pin

P18	17	D2	21	SW1 (switch 1)
P19	18	D3	22	SW2 (switch 2)
P20	19	D4	27	Speaker
P21	20	D5	28	Servo Port 1
P22	21	D6	29	Servo Port 2
P23	22	D7	30	Servo Port 3
P24	23	E0	8	I/O Expansion Port 1
P25	24	E1	9	I/O Expansion Port 2
P26	25	E2	10	I/O Expansion Port 3
P27 / AX4	26	A5	7	A/D port 4
P28 / AX0	13	A0	2	Potentiometer
P29 / AX1	14	A1	3	A/D port 1
P30 / AX2	28	A2	4	A/D port 2
P31 / AX3	27	A3	5	A/D port 3
Vdd	37	Vdd	11, 32	Positive Supply (5v) for Logic and I/O Pins
Vss	4, 39	Vss	12, 31	Ground Reference for Logic and I/O Pins
Vin	40	N/A	N/A	Power Input (9-12v) Feeds Vdd thru 5v Regulator
Res	38	MCLR	1	Reset Switch
S_in	2	A4	6	Atom Programming Pin thru 9 pin connector
S_out	1	A4	6	Atom Programming Pin thru 9 pin connector

ATN	3	MCLR	1	Atom Programming Pin thru 9 pin connector

Chapter 2 – Atom Basic Compiler

I briefly covered what a BASIC compiler is and how the Atom micro works. Now let's look at the Atom Basic Compiler software used to program the Atom modules. The Atom BASIC compiler includes many commands and the list is included in the next chapter with a brief description of each command. The Atom command manual has much more detail about each command and also several other details about using the Atom BASIC compiler. This book focuses on the projects at the end to teach you how to use the Atom to supplement the Atom manual not replace it.

Getting started programming the Atom does not require you to know every detail about every command. In fact I look up commands in the manual from time to time just because I forgot all the little details of the command setup. Just knowing that a command exists and generally what it does is all you really need to know. Then when you begin a new project you have a good idea of how your software will be structured and how you will attack the design.

Some of the Atom commands are incredibly simple and just turn an I/O port on or off. Others access the advanced features of the PIC such as internal Analog to Digital (A/D) converter and internal EEPROM. Still others are even more advanced and address the internal Microchip PIC hardware features such as interrupts and hardware serial port.

Before all this though, I want to cover the structure of an Atom program and the way to establish variables, call I/O, and some hardware layout details you will want to know. This format is similar to the Atom command manual but in my own words based on my experience with a few tricks thrown in that I've learned.

Memory

There are three types of memory you will work with in the Atom: RAM, EEPROM and Program Memory. All of these are contained inside the Microchip PIC microcontroller, which as I explained earlier, is the heart of the Atom micro.

RAM is random access memory; this type of memory is used to store variable values, and system values. RAM is also used to store the return location of GOSUB statements. RAM is temporary memory so when the Atom is powered down, this memory may erase. You cannot count on it erasing itself properly though because at start-up it can scramble and store strange results in your variable locations. This is why you should clear all RAM at the beginning of a program so you know its state. The CLEAR command mentioned later takes care of this for you.

EEPROM is on-chip data storage also known as non-volatile memory. This is commonly used to store values that will remain even when the Atom is shut down. In fact, the data will stay for up to 10 years according to Microchip, the PIC chip manufacturer. The EEPROM stores data in byte size. It can store 256 bytes.

Program Memory is the actual memory where your program will reside. The size of available program memory will limit the size of your program. The more complicated the program is, the more memory you will need. The Atom has 8K of memory. The first 2k gets used up pretty quick because the Atom compiler loads a bunch of command information first. After that, the program memory will get used much more efficiently. The Atom Basic compiler will tell you how much memory your program used at the bottom of the programming software IDE screen. For example, the simple program below flashes an LED one second on and one second off over and over again.

```
main
high 0              'Turn LED on
pause 1000          'Delay 1 second
low 0               'Turn LED off
pause 1000          'Delay 1 second
goto main           'Jump back to the top
```

When this program is compiled, it reports the following:

Program Memory Bytes Used(Library):   612
Program Memory Bytes Used(Tokens):   28
Program Memory Bytes Used(Total):   640
Program Memory Bytes Free: 13220

If you were to add those up, you would get a total of 14,500 bytes. "But wait, I thought you said the Atom only has 8000 (8k) bytes of program memory?" This is one of the more confusing aspects of using an Atom and it's because of the Microchip PIC. You see the PIC actually has a 14 bit wide memory, which is larger than a byte (8-bits) and smaller than a word (16-bits). Microchip, being a conservative company claims it as bytes. Now if you were to actually use the extra six bits for additional command data, you can get more memory out of the PIC than 8k. The Atom does this somewhat so they report 14k bytes of program memory. Officially the Atom has 8k.

In the end, all that matters is if your program fits or not, but many people want to jump from Basic Stamps or other PIC based modules and want to know if the program for those modules will fit in the Atom. There is no straight answer. It all depends on which commands were used and how they were used. Most Basic Stamp programs are less than 2k so they fit easily.

Built-in Hardware

Built-in hardware refers to additional circuitry that is built into the Microchip PIC. Examples of built-in hardware are Analog to Digital converters (A/D), Pulse Width Modulators (PWM), Hardware Asynchronous Receive/Transmit Serial ports (UARTS), independent timers/counters and so on. Built-in hardware adds pseudo multi-tasking abilities to the Atom, because in most cases it can be setup in your program and left to run while the main program does other things.

These functions can easily be controlled with Atom Basic commands but they are a bit advanced and it's best to study the Microchip PIC 16F876 data sheet before trying these out. Projects at the back section of this book show you how to use some of these functions.

Binary, Decimal and Hexadecimal Numbers

Before moving on, I wanted to briefly cover a topic you will need to understand to program Atoms successfully; numbers. I know you learned about numbers in kindergarten but to successfully program a microcontroller you need to know more than just the decimal system we learned early in life.

We normally use the characters 0123456789, this is the "decimal" or "base 10" system. Computers don't understand decimal numbers. They "know" only two states: on/off, yes/no, high/low, 1/0. The smallest unit of information they can store is a "bit", short for "binary digit". One bit can store one state: 1 or 0 (On or Off).
Computers count higher than 1 the same way we do; they group the digits to make larger numbers. In base 10, 9+1=10. In base 2 (also known as "binary"), 1+1=10. (That's "one zero", not "ten").

Binary numbers get long in a hurry. For example, the decimal number 201 is 11001001 in binary. Binary has unique carry points where it adds another column of digit. The first column is added at the 0 to 1 decimal transition. The second is added at the 1 to 2 decimal transitions. The next is added at the 3 to 4 and then 7 to 8, etc.

The table below shows the binary value for decimal 201 and the transition number above each digit or bit.

Decimal transitions

128	64	32	16	8	4	2	1
1	1	0	0	1	0	0	1

Binary Value

Every time a "1" appears, we include the decimal transition value above it in a mathematical equation to convert the binary number into a decimal number. We ignore the zeros. Therefore the block above equals:

$$128 + 64 + 8 + 1 = 201$$

Because it gets real tedious writing all those 1's and 0's as the numbers get bigger, we need a shorthand way of writing numbers that convert into binary much easier than decimal. This is where Hexadecimal numbers enter the picture.

Notice that the first 4 binary digits (8,4,2,1) represent 16 decimal values (0 thru 15). If we break a binary number into blocks of 4 bits, we can represent each block as a hexadecimal number, which is a base 16 systems. So, how do we represent the 16 combinations of a block with simple single digit numbers when only 10 single digit numbers (0-9) are available? We turn to the alphabet to make up the difference. We need 6 more characters, so we use A-F with A = 10 decimal, B = 11 decimal, C = 12 decimal, D = 13 decimal, E = 14 decimal and F = 15 decimal.

Using the 201 decimal example again and breaking it into blocks of 4-bits we get

1100 binary =12 decimal ="C" hex
1001 binary = 9 decimal = "9" hex

So 201 decimal = C9 hex and 11001001 binary

8	4	2	1	8	4	2	1
1	1	0	0	1	0	0	1
C				9			

Since Atom programs can accept binary, decimal and hexadecimal numbers and since we are using the same characters (1 and 0) in the various bases, we need some way to tell them apart in our Atom programs. For example is 1101 in binary, decimal, or hex?

Binary is represented by adding a "b" to the end or "%" to the beginning; 1101b, %1101.

Decimal is either a "d" or is not specified.
201d, 201

Hex is designated by adding a $ to the front or 0x.
$C9, 0xC9.

Most people use the "$" form. And yes, $FB00 is the same as $0000FB00. Also, note that hex digits mostly come in pairs. A pair of digits is known as a "byte", and is treated as a single unit.

So why do you need to know all this?
Because when you try to control an I/O port, in which each bit determines if the output is high or low, which is easier to understand?

Portb = 170

Portb = %10101111
Portb = $AF

To me, binary makes it easier to see that the last four bits of the port are set to a "1" or high. Also since I know "F" in hex is 1111, it's also pretty easy to tell. But 170 doesn't tell me a thing. This is why I spent the time to explain numbers.

Variables

In almost every program you write there will be a need to store some temporary information such as the result of a math operation or the value of an I/O port that was read. The Atom offers about 270 bytes of the PICs 368 bytes of RAM for this temporary storage. In order to use that memory in your Atom program, the program needs to declare how much space and what reference name your program will want for that memory. You do that with a variable declaration using the "var" directive.

A variable can point to different size blocks of RAM. It can point to a single bit or it can point to multiple bytes. To let the Atom module know how much memory your program wants to use with each variable name, you declare the size of the variable at the same time you establish a new variable name. For example, lets say you wanted to create a variable named "result" and want it to be able to store a value from 0 to 250. Since a byte can store up to 255 different values, we establish "result" as a byte variable with the following line:

```
result var byte  'establish byte size variable result
```

There are rules for variables.
1. Variable names must start with a letter.

2. They can contain letters, numbers and special characters, however they cannot be the same name as an Atom command or label used in a program.
3. The same variable name cannot be defined twice with the Var directive. The Atom does not distinguish between upper and lower case, so the name TVAR is equivalent to tvar.
4. The maximum character length is 1024 characters.
5. Variables can be defined as a Bit, Nibble, Byte, Word or Long (which is 4 bytes or 2 words)

Some examples of defined variables:

```
DOG Var Bit      ;0 to 1
POST Var Nib     ;0 to 15
LOG Var Byte     ;0 to 255
STICK Var Word   ;0 to 65535
TREE Var Long    ;0 to 4,294,967,295
```

Two more options are "sbyte" and "sword" which represent signed byte and signed word (positive and negative numbers).
Examples:

```
SLOG var SByte   ;-127 to +128
STREE var Sword ; -32767 to +32768
```

Some more tips on assigning sizes to Variables:

1. When assigning sizes to variables, keep in mind what the variable is being used for. For example, storing the ASCII value of the letter A, will require a Byte (0 to 255).
2. A variable should be the smallest size to hold the largest value that will be stored. If a variable will only hold the High / Low transition of an input pin (1 or 0), use a bit.

3. If a variable exceeds its maximum size, the excess bits will be truncated. If you were to load the binary value %11110000 into a NIB (nibble) sized variable the %1111 part would be lost.

Aliases

Aliases are alternate names for defined variables. As an example:

```
DOG Var Byte    ;DOG is assigned as an 8 bit variable (Byte)
CAT Var DOG     ;CAT now points to the variable DOG
```

In the above example if DOG were equal to 10, any time the variable CAT was accessed it would equal 10 also since it points to the same RAM location.

Aliases are a good idea when you want to reuse the same RAM space but in different functions.

Variable Modifiers

Variable modifiers are used to access only parts of a variable. An example would be if you had the word value of %0111111100000001 but only required access to the high byte of the word. You could then use the ".highbyte" modifier on the variable DOG with an alias CAT as shown below:

```
Dog Var Word
Cat Var Dog.HighByte
```

If the binary value of %0111111100000001 were written to Dog, any actions to the variable Cat would act on the high byte %01111111. If the LowByte modifier were used then CAT would only act on the %00000001 portion of DOG.

The following is a description of the Word and Byte sized modifiers.

HighByte: Left 8-bits of a word
Lowbyte: Right 8-bits of a word
HighNib: Left 4-bits of a byte
LowNib: Right 4-bits of a byte
HighBit: Always the highest bit
LowBit: Always the lowest bit (same as bit0)

The table below shows the rest of the different modifiers that can be used:

BIT0:   bit 0 of variable
BIT1:   bit 1 of variable
BIT2:   bit 2 of variable
BIT3:   bit 3 of variable
BIT4:   bit 4 of variable
BIT5:   bit 5 of variable
BIT6:   bit 6 of variable
BIT7:   bit 7 of variable
BIT8:   bit 8 of variable
BIT9:   bit 9 of variable
BIT10:   bit 10 of variable
BIT11:   bit 11 of variable
BIT12:   bit 12 of variable
BIT13:   bit 13 of variable
BIT14:   bit 14 of variable
BIT15:   bit 15 of variable
BIT16:   bit 16 of variable
BIT17:   bit 17 of variable
BIT18:   bit 18 of variable
BIT19:   bit 19 of variable
BIT20:   bit 20 of variable
BIT21:   bit 21 of variable
BIT22:   bit 22 of variable

BIT23:   bit 23 of variable
BIT24:   bit 24 of variable
BIT25:   bit 25 of variable
BIT26:   bit 26 of variable
BIT27:   bit 27 of variable
BIT28:   bit 28 of variable
BIT29:   bit 29 of variable
BIT30:   bit 30 of variable
BIT31:   bit 31 of variable

NIB0: nibble 0 of variable
NIB1: nibble 1 of variable
NIB2: nibble 2 of variable
NIB3: nibble 3 of variable
NIB4: nibble 4 of variable
NIB5: nibble 5 of variable
NIB6: nibble 6 of variable
NIB7: nibble 7 of variable

BYTE0:   byte 0 of variable
BYTE1:   byte 1 of variable
BYTE2:   byte 2 of variable
BYTE3:   byte 3 of variable

WORD0:   First 16 bytes
WORD1:   Last 16 bytes

MSBs and LSBs

Another often referred to term is LSB or MSB. These terms mean Least Significant Bit (LSB) and Most Significant Bit (MSB). In a byte value of %11110000, the MSB is 1 and the LSB is 0. MSB is the first bit on the left. LSB is the first bit on the right.

This is important to know if you serial shift data one bit at a time between Atom modules or external support circuitry. Did the data shift to the right or the left? It's easy to clarify when you state MSB first which means the first bit received is the left most bit or most significant bit.

Arrays

Another option is to declare a series of bytes that form an array. This is a single named variable with several sub-labels. As your program begins to perform more complex tasks, there will be times when you want a variable to hold many values. For example, you can create an array that can hold five different values of the same type as shown below:

Temp var Word(5)

The number "5" in parenthesis shows the variable temp has five cells (0 thru 4). Once the array has been defined, each cell can be accessed by its number:

Temp(0) = 10
Temp(1) = 25
Temp(2) = 45
Temp(3) = 55
Temp(4) = 65

The above will assign the value of 10 to the first cell in the 5-cell array, 25 to the second cell and so on. Using arrays can simplify your program as shown below:

```
Temp Var Byte(5)              ;variable temp now has 5 cells
Counter Var Byte

For Counter = 0 to 4          ;increment each cell
Temp(Counter) = Count +2   ' Each cell has a unique value
```

Next

The above code example will load each array, 0 to 4 (5 cells) with the array number + 2. Which if done manually would equal the following:

Temp(0) = 2
Temp(1) = 3
Temp(2) = 4
Temp(3) = 5
Temp(4) = 6

ASCII

ASCII is the numerical representation of the letter and characters we type on a computer. Each letter has a numeric value associated with it. This allows the microprocessor to manipulate text as numeric data, which is required because as I mentioned before, micro's work on 1's and 0's only. The Atom is no different. It will automatically convert any text between quotes into its respective ASCII values. This is useful when you are trying to work with the literal value of characters.

Example:
SEROUT 1,I9600,["A"]

The above code example will transmit the ASCII value of the character capital A in quotes. Numbers will also have an ASCII value associated with them, which is different than their own numerical value. For example, the number "1" has an ASCII value of 31. When you send numbers or letters to a PC or another module using serial RS232 type communication, it will be sent as a series of ASCII codes.

The ASCII chart is included at the back of this book for reference. It's handy to have because even common components such as LCD's use ASCII values in its character RAM.

Because ASCII is numeric, you can store strings of characters and words in a single variable or variable array. This is handy for special messages you may want to display on an LCD. Just call up the variable contents and send it to the display device.

Line Labels

Line labels are not variables but it seems like they should be explained at the same time. Line labels are just marking points in a program that allow you to jump from a section of code to another section of code easily.

Original versions of the Basic language had a line number for each command line but Atom does not use line numbers. All the Atom program needs is an occasional line label so the command GOTO or If-Then knows where you want to jump to when you use those commands.

As an example:
Loop:
    High 0
    Pause 1000
    Low 0
    Pause 1000
Goto Loop

This program above will run in a continuous loop because when the program gets to the GOTO line it jumps to the "Loop:" line and starts it all over again. In fact all your Atom programs should loop like this otherwise when the last command was run, it will just lock up and stay that way until power is removed. You don't want that to happen.

Tables

Tables are another way of using memory to store information used in the Atom program. Tables are made up of constant values, which can be byte, word or long size.

One example of tables would be if you had a program that sent data to an LCD screen. Suppose your program was setup to send data to the LCD in a menu format and the menu format was made up of many different text messages. Instead of copying the text strings over and over wherever they are needed in the program, you could use a table as a single storage point that your program jumps to when it needs the data.

A table acts a lot like a line label that I just described except it points to the start of a block of data rather than the start of a block of code. A table is setup similar to a variable. It has a label and a size directive. Bytetable is used to tell the Atom program that the data should be stored in the table as a series of bytes.

The Table format is shown below:

Label TableType Data, Data, ...Data

Label - is the name of the table used to call or access the table.

TableType - is the size of the table data. Tables can be ByteTable, WordTable, LongTable or FloatTable.

Data - is the constant values or constant expression stored in the table.

Table Code Example:

FirstMenu ByteTable "Enter An Option"

LCDWRITE 7\5\6, OUTA, [str firstmenu\15]

This isn't a complete program but it shows a byte size table setup with a series of characters that form a sentence to be displayed on an LCD. Because the data is in quotes, the Atom compiler treats it as a series of ASCII characters each one byte long. When the LCDWRITE command (to be explained later) sends the sentence to the LCD, it sends the 15 bytes to the LCD by reading each ASCII character one at a time from the byte table.

Notice how there is really only 13 letters stored in the byte table but the LCDWRITE command sends 15 ASCII characters. This is because the space between "Enter" and "An" and the space between "An" and "Option" also has an ASCII value associated with them. Therefore there are 15 ASCII characters stored at the "FirstMenu" byte table.

Constants

Sometimes you may want to declare a fixed value throughout your program but you want the ability to change it easily in future applications of the program. A constant is a way to declare a value at one location and tie it to a nickname, similar to a variable. This allows you to change the constant value in one location and thus changing it throughout the entire program at the same time.

The format for declaring a constant is similar to variable but uses the "CON" directive instead of "VAR". Here are some examples:

```
Meter CON 1                'Meter is a constant = 1
Centimeter CON 100         'Centimeter is a constant = 100
Millimeter CON 1000        'Millimeter is a constant = 1000
```

Constants can also make it easier to describe what the constant is for and can make your program easier to understand. Another good use of the constant is when you have values that are based on other values. You can add a math modifier to the constant when it's declared.
For example:

meter CON 1   'Meter is a constant = 1
centimeter CON meter * 100   'Centimeter is a constant = 100
millimeter CON centimeter * 10 'Millimeter is a constant = 1000

In the above example "centimeter" and "millimeter" values were derived from the constant "meter". Now you place all these constants in your program where you need it and then later if "meter" has to change, then by just changing the value associated with "meter" you will automatically change all the "centimeter" and "millimeter" constants at the same time.

The Atom pin names are also considered to be constants so they can be renamed to the function they are connected to by setting them up at the beginning of your program. For example:

RedLed Con P0   ' Red LED is connected to P0 pin
GreenLed Con 1   ' green LED is connected to P1 pin

Main
High RedLed   ' Use the constant names instead of the pin names
High GreenLed   '   throughout the program
Goto Main

It's easy to see that this method allows you to easily change your hardware, such as moving the red LED to pin P2 by just changing the constant line to "RedLed Con P2" thus changing it throughout the whole program. The more complex your program is the more useful this becomes.

One thing to consider is constants will not change so they should not be used in comparison expressions like below where you may be monitoring the state of an I/O pin:

Tpin Con IN1

Input Tpin
If Tpin = 0 Then SomeLabel

The If...Then statement will always be true since constants will not change during the running of the program. The correct way to perform a true / false statement on an input pin is to treat it as a variable as shown below:
Tpin Var IN1

Input Tpin
If Tpin = 0 Then SomeLabel
Tpin can now change state within the program because variables can change. In this case , because the variable is actually an I/O pin, the variable doesn't point to a temporary memory location but instead points to the port data register, which I will talk more about in the next section.

Inputs/Outputs (I/O)
In my opinion this is the most important section of this book for the beginner. Everything the Atom module does (and the Microchip PIC inside of it) has to go through the I/O pins. In fact, setting I/O pins high (1 or 5volts) or low (0 or ground) is essentially all the Atom module does, or any microcontroller for that matter. Serial communication, controlling a complex industrial system, flashing a simple LED, they all are based on turning an output pin high or low at a controlled pace.

It's not just as simple as it sounds though because the I/O pins have multiple functions. They can be an output; setting pins high or low or they can be an input, reading a high or low voltage level. Some pins (analog to digital converter) can even read a variable voltage level that gets converted into a series of high and low signals or bits.

Inside the Atom micro each I/O pin is directly connected to an I/O pin of the internal Microchip PIC microcontroller. A few have some special circuitry between them such as the programming pins but most of the I/O is directly connected. Therefore all I/O control requires the same actions required to setup the Microchip PIC microcontroller. Therefore, to fully understand all the workings of the Atom I/O it's best to understand the Microchip PIC I/O. I highly recommend that you read the Microchip data sheet for the PIC16F87x or PIC16F88x parts which you can download from microchip.com. Beyond the data sheet though, I wanted to cover some of the Microchip PIC I/O functions to help you understand the Atom.

The PIC I/O is setup in several blocks of eight pins also called a port. A port is eight bits wide or a byte wide and each pin is controlled by a bit in the port control byte. The Atom modules that use the 28 pin interpreter chip have ports A, B and C while the 40 pin interpreter chip modules also add ports D and E. Each port has at least two control registers (special memory locations), a data register and a direction register. All the PIC/Atom I/O is controlled by these two registers so it's best to really understand these.

Data Direction Register
The way to tell the PIC/Atom I/O pin to function as an Input or Output requires you to set up the Data Direction Register first. The data direction register is actually called the TRIS register or TRI-State register in the Microchip data sheet. Tri means three and the three states are High, Low and High Impedance. Each I/O port has its own TRIS register named after the port letter. TRISA is for PortA, TRISB for PortB, etc.

Each pin of the port can be setup as an input or an output by controlling a bit within the TRIS register. A "1" makes the pin an input and a "0" makes the pin an output. You can arrange these ports in any combination of input and output. You can change the direction at any point in your program. With proper care, you can read a port as an input and then drive something from the same port as an output.

The Atom compiler allows you to set this directly by directly declaring it:

TrisB = %11110000   'P7-P4 are inputs, P3-P0 are outputs

This only sets the direction though. A second register, data register, is where the actual state of the I/O pins is stored.

Data Register
If your program reads from or writes to a port or just a pin it will actually be reading from or writing to the data register which goes by the port's name such as PortB. You can manipulate the port directly the same way the Tris register was setup.

Portb = %11110000   'P7-P4 are high, P3-P0 are low

You can also read the port directly.

Value = portb

You can even read individual port pins with a suffix.

Value = portb.bit1

Other Methods of I/O control

Now the the Atom compiler also carries over some popular declarations that other compiler software uses such as DIRS, INS and OUTS to make all this easier to control the I/O but recognize that all these are just different ways of doing the same thing, controlling the TRIS register and reading from or writing to the PORT register.

Here are some of the other accepted ways to access and control the Atom module I/O.

Data Direction Register Control

DIRS plus 16-bit value controls TRISB and TRISC registers for P0-P15.
DIRL plus 8-bit value controls the TRISB register for P0-P7.
DIRH plus 8-bit value controls the TRISC register for P8-P15.
DIRA plus 4-bit value controls the lower 4 bits of TRISB for P0-P3.
DIRB plus 4-bit value controls the upper 4 bits of TRISB for P4-P7.
DIRC plus 4-bit value controls the lower 4 bits of TRISC for P8-P11.
DIRD plus 4-bit value controls the upper 4 bits of TRISC for P12-P15.
DIR# (# is any number from 0 to 31) accesses the direction of P0 - P31 individually.

Reading from the port

INS receives a 16-bit value read from PortB and PortC registers for P0-P15.
INL receives an 8-bit value read from PortB register for P0-P7.
INH receives an 8-bit value read from PortC register for P8-P15.
INA receives a 4-bit value read from PortB register for P0-P3.

INB receives a 4-bit value read from PortB register for P4-P7.
INC receives a 4-bit value read from PortC register for P8-P11.
IND receives a 4-bit value read from PortC register for P12-P15.
IN# (# is any number from 0 to 31) receives the state of P0 - P31 individually.

Writing to the port
OUTS sends a 16-bit value to PortB and PortC registers for P0-P15.
OUTL sends an 8-bit value to PortB register for P0-P7.
OUTH sends an 8-bit value to PortC register for P8-P15.
OUTA sends a 4-bit value to PortB register for P0-P3.
OUTB a sends a 4-bit value to PortB register for P4-P7.
OUTC sends a 4-bit value to PortC register for P8-P11.
OUTD sends a 4-bit value to PortC register for P12-P15.
OUT# (# is any number from 0 to 31) sets the state of P0 - P31 individually.

Important Note!
Some of these are slightly different than what other modules do such as the Basic Stamp modules. When the Atom uses the DIRS = $FF to set all the pins of the PortB and Port C registers to inputs, it does it by making those control bits a "1" just like the TRIS register requires. The Basic Stamp is the opposite, where it's DIRS command makes the pins inputs by clearing the bits to "0". This can cause confusion if you try to convert a Basic Stamp program to an Atom program. This is why I say this section is so important. Understanding how the internal control of the PIC's I/O works makes you a better programmer and makes your programs more reliable.

Basic Command control of I/O
Most of the Atom compiler commands automatically setup the data direction register and control the data register for you which is a reason programming in Atom Basic makes controlling a microcontroller so easy.

Commands such High and Low automatically set the proper data direction register for the I/O pin it is working on and then change the state of the bit in the data register. Therefore all that stuff I covered about how to control the I/O isn't needed in most cases but if you really want to write good reliable programs, you have to understand how the I/O works.

There are even Atom Basic commands that just operate on the Tris register such as INPUT and OUTPUT. These are just simple ways of set or clearing the port bit in the TRIS register.

INPUT 0   ' Sets Pin 0 as an Input
OUTPUT 1 ' Sets Pin1 to an Output

Pin Names
The Atom I/O pins have multiple ways they can be called.
They typically are called by the letter P followed by their pin number such as P1. They can also just be called by the number alone like "HIGH 1" sets pin P1 high. Pins can also be called by their port and pin number in the following format; portb.bit0 refers to pin P0. This can be a little confusing because the numbers don't always match. For example, P8 is the same as portc.bit0. Knowing how the Atom module circuitry is laid out in the Atom module is very helpful.

Atom pins can also be directly designated as an input by adding "IN" in front of the pin number. IN0 refers to the input state of the P0 pin. This is used often in IF-THEN statements such as:

If in0 = 1 then jump   'Move to jump label is pin P0 is high

Analog to Digital (A/D) pins

Some of the Atom module pins are connected to the PIC's internal A/D converter. These pins will sometimes have a unique name such as AX0, AX1, AX2 and AX3. This was Basic Micro's way of distinguishing them from the standard I/O pins. The problem is these pins can also be setup as digital I/O pins.

On the 40 pin Atom interpreter chip, the Atom also has additional A/D pins that are actually on Port E. These are pins P24 thru P26. These were not given AX names even though they are A/D pins. You can use their P names in the ADIN command line. The P27 pin is another additional A/D pin. It's connected to PortA bit 5. The P27 pin has an AX name, AX4. You can use P27 or AX4 in the ADIN command. This is only available on the 40 pin interpreter chip based Atom module or BasicBoard.

Here is the list of eight A/D pins available on the various Atom modules:

<u>Atom 40 and BasicBoard</u>
P24   A/D pin (40 pin interpreter chip only)
P25   A/D pin (40 pin interpreter chip only)
P26   A/D pin (40 pin interpreter chip only)
P27/AX4   A/D pin (40 pin interpreter chip only)
P28/AX0   A/D pin
P29/AX1   A/D pin
P30/AX2   A/D pin (Vref-)
P31/AX3   A/D pin (Vref+)

<u>Atom 28 and Ultimate OEM</u>
AX0   A/D pin
AX1   A/D pin
AX2   A/D pin (Vref-)
AX3   A/D pin (Vref+)

<u>Atom 24 and Atom OEM</u>
AX0 A/D pin (Vref-)
AX1 A/D pin (Vref+)
AX3 A/D pin

Note:
AX2 is not an A/D pin on these modules. It is just an open drain output
and also an input pin to the Timer0 register.

Inside the Analog to Digital Converter

Inside the Atom micro interpreter chip are several registers that setup the
A/D for proper operation. All this is easily done for you with a single
ADIN Atom Basic command but I wanted to describe what it's doing so
you can understand the A/D ports better.

The A/D structure in the PIC uses four registers for access, control and
value storage. The "A/D control register 0" (ADCON0), the "A/D control
register 1" (ADCON1) and the "A/D result registers" (ADRESL and
ADRESH). The ADCON0 register is really more of a control register
while ADCON1 is a setup register.

ADCON0

ADCS1	ADCS0	CHS2	CHS1	CHS0	GO/$\overline{\text{DONE}}$	Not Used	ADON
Bit 7	Bit 6	Bit 5	Bit 4	Bit 3	Bit 2	Bit 1	Bit 0

ADCS1,0: A/D Conversion Clock Select
00 - External Oscillator / 2
01 - External Oscillator / 8
10 - External Oscillator / 32
11 – Internal RC Oscillator

The ADCS1 and ADCS0 two bits allow you to pick from four different clock sources. The clock source is used in the sample and hold A/D circuitry inside the PIC.

The Atom Basic ADIN command controls these two bits with the "clk" selection in the command line. The simplest choice is the internal RC oscillator (clk = 3) since it runs independent of the external crystal/resonator. The other choices are for more precise measurements.

According to the PIC 16F876 data sheet though, the 00 (oscillator/2) and 01 (oscillator/8) selections are not accurate if the PIC's external clock speed is above 5 Mhz. Since the Atom runs at 20 Mhz, I recommend you use the 10 (clk/32) or 11 (RC) options, which are the "clk" option 2 and 3 in the Atom ADIN command.

CHS2-0: Analog Channel Select

000 – Channel 0 (A0 pin)
001 – Channel 1 (A1 pin)
010 – Channel 2 (A2 pin)
011 – Channel 3 (A3 pin)
100 – Channel 4 (A5 pin)
101 – Channel 5 (E0 pin)
110 – Channel 6 (E1 pin)
111 – Channel 7 (E2 pin)

These bits choose which A/D port you want to read within your program. Atom Basic automatically selects this when you designate which "pin" to use in the ADIN command.

Notice how pin A4 is skipped. That is because A4 is used for the timer TMR0 input and also is an open source output. Any digital I/O on A4 requires an external pull-up resistor. This is the AX2 pin I talked about earlier, which is only available on the Atom 24 and OEM.

$\overline{GO/DONE}$ – A/D Conversion Status bit

This bit is really a control bit and an indicator flag. It is used to monitor when the A/D conversion is complete. It allows your program to check A/D status. When it is set to a "1", the A/D conversion process starts. This bit is automatically cleared when the conversion is complete. The ADIN command monitors this to determine when the conversion is complete.

Not Used bit

This bit is not used for anything and can be ignored.

ADON – A/D On bit

This bit turns the A/D circuitry on or off. Setting this bit to a "1" will enable the A/D converter at the channel selected in bits CHS2-0. Setting this bit to a "0" shuts down the A/D circuitry so it doesn't draw any current.

The ADON bit has to be set before the GO bit is set. In fact, the PIC requires your program to delay one sample time period between turning the A/D converter on (ADON = 1) and starting the conversion (GO = 1). That time has to be calculated to be exact but it's usually less than 50 usec. ADIN takes care of this for you.

ADCON1 Register

ADFM	Not Used	Not Used	Not Used	PCFG3	PCFG2	PCFG1	PCFG0

The ADCON1 register is where PortA and Port E are setup to be digital or A/D ports. Remember, the TRISA register just makes the port an input or output. ADCON1 takes it the next step and selects what kind of input. This is only used on PortA and PortE because they share standard I/O circuitry with the A/D converter circuitry. Once again, the ADIN command takes care of all this. It does give you some options though in the "adsetup" designator.

The ADCON1 bits are as follows:

ADFM – A/D Result Format
This bit selects how the 10-bit result is stored in a 16-bit variable space.
0 – Left Justified
1 – Right Justified

The "AD_LON" (left justified) and "AD_RON" (right justified) in the ADIN command control this bit.

PCFG3-0: A/D Port Configuration Control
These bits set which A/D port pins are non-A/D digital inputs and which are analog A/D converter pins. It also selects the voltage reference used by the A/D converter. The "AD_LPOS", "AD_LNEG", "AD_RPOS" and "AD_RNEG" in the ADIN command control some of these bits.

Within these bits are where the Vref- and Vref+ pins are setup. If you use these on the Atom, you are shrinking the voltage range for the A/D conversion. Normally the A/D conversion divides 5 volts by 1024 A/D values for 10-bit resolution. That works out to 4.8 millivolts per A/D bit. If you are measuring a small range of voltage such as 0 to 2.5v then you can increase the resolution to 2.4 millivolts per A/D bit. You do this by setting up the A/D register to use the VREF+ pin and then applying 2.5v reference on that pin using a resistor divider. It's really an advanced technique that gets deeper than the intentions of this book. I just wanted to cover it briefly so you were aware of this option.

ADRESH and ADRESL
These registers are where the result of the A/D conversion is stored. The

A/D result stored in these registers is transferred to the word byte you designate in the ADIN command line.

This should give you a brief but effective understanding of the Atom micro (PIC) A/D port. The Chapter 11 project is an example of using the A/D.

Internal Timers

Another special function inside the Atom micro is the timers. The PIC16F876 and 877 Atom chips have three timer modules. But what is a timer, what does it do and how do we use in with Atom? Those are question we plan to answer with a project towards the end of this book and the description below.

The so-called "Timer" inside of a Microchip PIC is really just a binary counter circuit that is fed by a controlled frequency clock source. In other words, it's not a stopwatch or clock giving you minutes, seconds or tenths of second's value to display somewhere that is meaningful to humans. It's just a reference counter with an accurate time base. Therefore we can use the value obtained from the timer peripherals in the PIC to calculate various functions, such as the time between one input changing to a high state and a second input changing to a high state. We can use it as a reference indicating when to change an output from low to high or high to low at a constant rate.

8-Bit Timer

bit 7	bit 6	bit 5	bit 4	bit 3	bit 2	bit 1	bit 0	
0	0	0	0	0	0	0	0	← ⊓⊔⊓⊔⊓⊔
0	0	0	0	0	0	0	1	Clock Pulse 1
0	0	0	0	0	0	1	0	Clock Pulse 2
0	0	0	0	0	0	1	1	Clock Pulse 3
⋮	⋮	⋮	⋮	⋮	⋮	⋮	⋮	
1	1	1	1	1	1	1	1	Clock Pulse 256

PIC timers come in three variations and have three different names, TMR0, TMR1 and TMR2. Two are 8-bit (TMR0 and TMR2) and one is 16-bit wide (TMR1). Because timers are binary counters, the 8-bit timers can count from 0 to 255 (binary 0 to binary 11111111) and the 16-bit timer can count from 0 to 65535 (binary 0 to binary 1111111111111111).

The three timers have different features that make them unique and useful for different applications.

TMR0
- 8-Bit Timer
- Readable and writeable as one byte
- Can be fed from internal clock or external input pin (RA4)
- Can be set to create a hardware interrupt at overflow (255 > 0)
- 8-bit prescaler 1:2 to 1:256
- Edge selectable for external input

TMR1
- 16-Bit Timer
- Readable and writeable as two bytes
- Can be fed from internal clock or external clock crystal
- Can be set to create a hardware interrupt at overflow (65535 > 0)
- 4-bit prescaler 1:2 to 1:8

TMR2
- 8-Bit Timer
- Readable and writeable as one byte
- Writable comparison byte size register
- Only fed from internal clock
- Constantly compared to secondary presettable binary value
- Can have 1:1, 1:4, 1:16 prescaler or 1:1, 1:2, 1:3 to 1:16 postscaler
- Output can drive synchronous port
- Can be set to create a hardware interrupt at match of preset binary value

All three can be fed from the internal PIC clock but TMR0 can also be fed from an external input pin. This allows TMR0 to act as an event counter rather than a timer. TMR1 can be controlled by an external crystal separate from the internal PIC clock or from an external input making it a 16-bit counter. This offers the opportunity to control TMR1 externally from a slower clock source such as a digital watch crystal or a digital counter source. TMR2 can only run from the internal PIC clock but can be automatically set to constantly check if it's reached a preset value similar to a cooking timer.

The table below shows the features of the three timers along with the control bits that setup these features.

Features	TMR0	TMR1	TMR2

73

Size	8-Bit	16-Bit	8-Bit
Prescaler	OPTION_REG.3 - 0 %1xxx = 1:1 %0000 = 1:2 %0001 = 1:4 : %0111 = 1:256	T1CON.5 - T1CON.4 %00 = 1:1 %01 = 1:2 %10 = 1:4 %11 = 1:16	T2CON.1 - T2CON.0 %00 = 1 %01 = 4 %1x = 16
Postscaler	Not Available	Not Available	T2CON.6 - 3 %0000 = 1:1 : %1111 = 1:16
Interrupt Enable Bit	INTCON.5	PIE1.0 and INTCON.6	PIE1.1 and INTCON.6
Interrupt Flag	INTCON.2	PIR.0	PIR.1
Internal Clock	Fosc/4 Selected by OPTION_REG.5 = 0	Fosc/4 Selected by T1CON.1 = 0	Fosc/4 (Only Option)
External Crystal/Resonator	Not Available	Crystal or Resonator connected between C0 and C1 pins Selected by T1CON.1 = 1 and T1CON.3 = 1 Sync external with internal clock selected by T1CON.2 0 = Synchronize 1 = Do Not Sync	Not Available

Counter Mode or External Clock Mode	Pulse signal connected to TOCKI Pin Selected by OPTION_REG.5 = 1	Pulse signal connected to C0 pin Selected by T1CON.1 = 1 and T1CON.3 = 1	Not Available
	Edge Select Bit for Incrementing: OPTION_REG.4 0 = Low to High 1 = High to Low	Sync external with internal clock selected by T1CON.2 0 = Synchronize 1 = Do Not Sync	
On/Off Control	Not Available (Always on)	T1CON.0 0 = Off 1 = ON	T2CON.2 0 = Off 1 = ON
Timer Register Name(s)	TMR0	TMR1H – High Byte TMR1L – Low Byte	TMR2

Each of the timers has a register where the timer value can be read or written to. In the case of TMR1, which is 16-bits wide, it has two registers because the PIC has an 8-bit wide data bus TMR1H and TMR1L. These are the high byte and low byte values. Combined they form a word. To access this timer's value you have to read each register separately and then combine them into a word variable.

Atom has several commands to make it easy to setup and read or write to these timers. You can also read and write to the registers directly as Atom has reserved the register names in its structure. For example, to preset TMR0 to 56 so it will overflow on the 200th pulse rather than 256, you just add the statement below to your code.

TMR0 = 56   ' Preset TMR0 to 56 or %00111000 binary

If you are running the TMR0 timer in counter mode, you can retrieve its value by just storing it in a variable with the statements below.

countervalue var byte   ' Setup byte variable named
"countervalue"

countervalue = TMR0   ' Store TMR0 value in variable
countervalue

Prescaler/Postscaler
In addition to the binary counter, timers also have a prescaler or postscaler attached to their input or output. A prescaler and postscaler are the same thing just one is at the input of the binary counter (prescaler) and the other is at the output (postscaler). A prescaler or postscaler is just a shift register, which has a selectable, via software, output position.

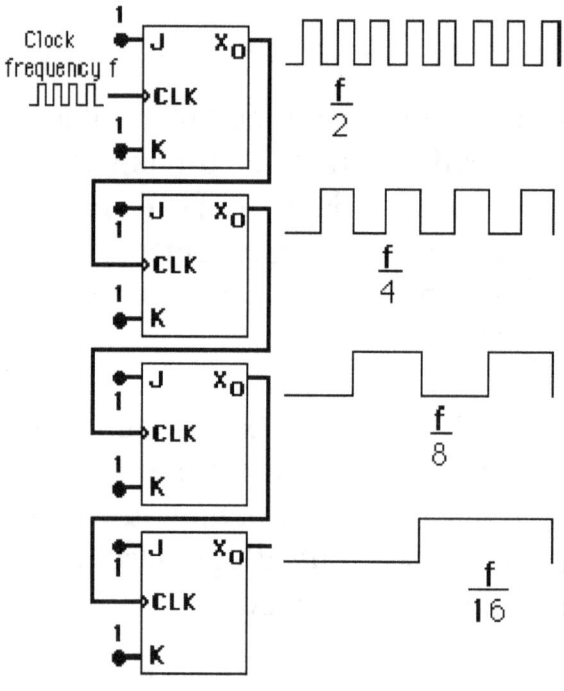

You can make them output every 2nd pulse, 4th pulse, 8th pulse, up to 256th pulse (prescaler) or every 1st, 2nd, 3rd, up to 16th pulse (postscaler). What these do is add a way to slow down the clock signal so the binary counter doesn't overflow or output a signal as quickly. If I have an internal PIC clock running at 4 Mhz, then running it thru a 1:4 prescaler will slow the clock down to 1 Mhz at the input to the timer's binary counter.

The postscaler, which is only available on the TMR2 timer, can make the timer hit its target value more than once before outputting a signal. If the postscaler is set to 1:4, the TMR2 has to hit the target value and output a signal to the postscaler four times before the postscaler sends a signal to the program, running in the PIC, can react to it.

Just like the A/D function, I recommend you read the Microchip PIC 16F876 or 877 data sheet about timers to fully understand their capability

Atom Math Functions
I won't get into great detail as the Atom manual covers this fairly well. The Atom Basic language will add, subtract, multiply and divide numbers with simple math statements. The Atom will also do floating point math which means it will work on numbers with a decimal point and give decimal point answers.
Atom Basic will also perform digital math such as AND, OR, Exclusive OR, etc. The Atom has sine and cosine functions. The Atom will also perform signed math, which means it can distinguish between positive and negative numbers.

Integer Math in general
The Basic Atom uses standard algebraic syntax.

$2+2*5/10 = 3$

The Atom will perform math in the proper mathematical order. The multiply and divide operators will happen first followed by the addition (and subtraction if that was in the equation).

In the above calculation 2*5 will be calculated first (equaling 10), then the divide by 10 (equals 1), then the addition of 2 (equaling 3).

Standard Math Commands

exp1 + exp2   Add exp1 to exp2
exp1 - exp2   Subtract exp2 from exp1

exp1 * exp2   Get lower 32 bits of multiply
exp1 ** exp2   Get high 32bits of a multiply
exp1 */ exp2   Get the middle 32 bits of multiply

exp1 / exp2   Get quotient of exp1 divided by exp2
exp1 // exp2   Get remainder only of exp1 divided by exp2

Multiplication

The Atom can perform 32x32 bit multiplication (i.e. 4,294,967,295 x 4,294,967,295 max.). If you use the maximum values only the lower 32 bits will be returned. That is why a second multiplication command is used. The '**' command allows you to get the high 32bits of data from the multiplication.
So as the example below shows, you must use two commands to get the full value from a large multiplication:

lowmult VAR long
highmult VAR long
middilemult VAR long

' ****** returns the LOW 32 bits of value = 5119617025
lowmult = 4,294,967,295 * 4,294,967,295

' ****** returns the HIGH 32 bits of value = 1844674406
highmult = 4,294,967,295 ** 4,294,967,295

'****** returns the MIDDLE word of value = 7440651196
middlemult = 4,294,967,295 */ 4,294,967,295

Division

The Atom can perform 32x32 bit division (i.e. 4,294,967,295 /
4,294,967,295 max.). The division command has two separate commands
associated with it.

The first command '/ ' will return the quotient value. Which is the actual
result from the divide. The second command "//", is used to return the
remainder if there is one.

For example:

divval VAR word
divrem VAR word

divval = 65000 / 101   ' Returns the Quotient value = 643
divrem = 65000 // 101   ' Returns the remainder value = 57

Signed Math

The Atom will handle negative numbers if asked to do so when the
variables are setup. This requires a special variable setup. Instead of
creating just a byte variable, you would create an Sbyte variable. What
this does is still use one byte of RAM space but uses the most significant

bit to indicate positive or negative number. You can also declare an Sword or a Long (4 bytes) for floating point.

An SBYTE is limited to values of +- 127. This is because the byte only has 7 bits for the value when it uses the 8th bit for the sign. An SWORD is limited to -32767 to +32768. BITs can't be negative, NIBs can't be negative. Longs are limited to -2147483647 to +2147483648

If you try to use negative numbers without declaring the variable as a Signed Byte or Word, then the result will not be accurate.

Other Math Functions
The following is a brief list of special math functions the Basic Atom can perform.

Unary Commands
variable = COMMAND value

ABS	Absolute value
SIN	sine of value(0-255)
COS	cosine of value(0-255)
DCD	2 to the nth power(n = value)
NCD	smallest power of 2 that is GREATER than value.
SQR	square root of value.
DEC2BCD	Convert integer to packed BCD format
BCD2DEC	Convert packed BCD to integer.

BINARY Commands
variable = value1 COMMAND value2

MAX - variable equals smaller of the two values.
MIN - variable equals larger of the two values.

DIG - variable equals the digit at the position designated by the value2 number

REV – variable equals value1 with the number of bits reversed designated by value2

Bitwise operators

variable = value1 COMMAND value2

& (AND)

variable = %01010101 & %00001111 'variable equals %0101

| (OR)

variable = %11110000 | %00001111   'variable = %11111111

^ (XOR)

variable = %10101010 ^ %11110000 'variable equals %01011010

\>> (SHIFTRIGHT)

variable = %11110000 >> 4   'variable equals %1111

<< (SHIFTLEFT)

variable = %00001111 << 4   'variable equals %11110000

Other Mathematical Functions

= Equal

<> Not Equal

< LessThan

\> GreaterThan

<= LessThan Equal

\>= GreaterThan or Equal

Example:

if value > 5 then jump   ' Test "value" to see if it's larger than 5

Floating Point Math

The Atom can perform 32x32 bit IEEE 754 Compliant Floating Point calculations with some minor variations. 32-bit floating-point math is CPU intensive since the Atom is only an 8-bit processor. Some special commands are required to handle floating-point math. No command can actually use floating-point math in its variable or expression. You will need to do some conversion with the floating-point value to send it (except when using the REAL modifier) using SEROUT or receive it using SERIN.

INT expression	Convert floating-point number to integer
FLOAT expression	Convert integer number to floating point
FNEG expression	Negates floating point numbers
exp1 FADD exp2	Adds two floating point numbers
exp1 FSUB exp2	Subtracts two floating point numbers
exp1 FMUL exp2	Multiplies two floating point numbers
exp1 FDIV exp2	Divides two floating point numbers

Floating Point Format

The floating-point math the Atom uses differs somewhat from the normal IEEE 754 floating point standard. The differences are minor. The next section gives a good indicator on how they differ.

Floating Point Example Program

```
temp var long
temp2 var long

Main
temp = 0            'Stores the integer value 0 in temp
temp2 = float 2     'Stores the floating point value 2.0 in temp2

temp = 2.1 FMUL temp2        ' temp now equals 4.2
```

Command Modifiers

Command modifiers can be used to modify the variable data within the Atom Basic command line. This is really just a shortcut to doing it all on separate lines.

Modifiers can be used with any commands that show {Modifier} in their syntax. An example would be if you used the SERIN command to receive a number from a PC serial port and store the value into a variable called "temp". The SERIN command will receive the ASCII value of the number and store the ASCII value in "temp". To have the SERIN command convert the value received into the decimal value before storing it in "temp" you just add the "DEC" modifier before the variable.

An example would be:
SERIN 0, i9600, [DEC TEMP]

If three ASCII characters 1,2 and 3 were received, then the DEC modifier would combine them into one value held in the variable "temp" as 123 (One Hundred and Twenty Three). If HEX was used as the modifier in place of DEC the value would be $123, if BIN was used the value would be %1111011. There are lots of modifiers including some that affect the I/O.

Command Modifiers

dec	Convert to Decimal Value
hex	Convert to Hexadecimal Value
bin	Convert to Binary Value
str	Convert string of values into an array of values

Signed Modifiers

sdec	Convert to signed Decimal Value
shex	Convert to signed Hexadecimal Value
sbin	Convert to signed Binary Value

Indicated I/O Modifiers – these modifiers convert a value to binary or hexadecimal then assigns an indicator "%" or "$".

ihex	Convert to Hexadecimal Value
ibin	Convert to Binary Value

Combination I/O Modifiers

ishex	Convert to Signed Hexadecimal Value
isbin	Convert to Signed Binary Value

Output Only Modifiers

rep	Repeat sending output character *n* times
real	Convert a Floating point value to ASCII characters. (Sign and decimal point are handled)

Input Only Modifiers
waitstr

waitstr var\count{\eol char} - Compares received characters to variable array,waiting until they are equal or until EOL (End of line) character is found. The EOL character will override the count value if both EOL character and count values are specified.

> Example:
> SERIN 1,i9600, [WAITSTR TEMP\10\"E"]   'compares values in
>   'an array until "E" is
>   ' received or 10
>   'characters have been
>   'received.

wait

wait (constant list) - Constant list can be "text" or 1,2,3,4,5(comma delimited). The wait modifier will wait for the specific characters.

Example:
    SERIN 1,i9600, [WAIT 1,"a",3,"c"]   'waits until 1, "a", 3,
                                          ' "c" has been
                                          ' received.

skip
Skip specified number of values.

Example:
    SERIN 1,i9600, [SKIP\10 TEMP]   'skips 10 characters
                                     'before loading
                                     'anything into the
                                     'variable temp.

Chapter 3 – Atom Commands

There is no reason to reproduce the whole Atom command manual since both the Atom Syntax manual and Atom Compiler manual can be downloaded for free from basicmicro.com. I do find it helpful though, to have an easy command reference to flip back to when you are trying to reproduce the project code in later chapters. Therefore, this chapter includes a brief listing of the Atom commands so you can easily check if a command exists and possibly help you with syntax errors. I list the commands in alphabetical order to make it easier to find when you search.

An exception to the alphabetical list is the advanced commands. I list those commands, which take advantage of the internal PIC special functions such as interrupts, hardware PWM and timers, at the end of this chapter under the title "advanced commands". The original Atom manual had it setup that way and I liked the format. I knew when I was looking for one of those advanced functions all the commands were in one place. If I listed them alphabetically then commands that reset the timer and others that setup the timer would be in different places in the alphabetical list. By placing them in their own category, the reader can find all the commands in one place for the timer or hardware serial port.

Some of this information is just the same wording as the original Basic Micro manual written by the Basic Micro engineering team that created the Atom modules. In other cases I described the command a little different where I thought I could explain it better. Let's start with the letter A and the ADIN command.

Standard Commands

ADIN pin,clk,adsetup, var
Sets up the Hardware A/D converter, and stores the value in a variable.

Pin - can be a constant or variable. The pin specifies which analog pin to use with the ADIN command.

Clk - can be a constant or variable. The Clk option sets the sampling time for the A/D conversion

Adsetup - can be a constant or variable. Adsetup is used to setup the options available with the ATOM hardware. The available options are shown below:

AD_LON Left justified. 6 Least Significant bits are read as '0'.
AD_LPOS Left justified with positive voltage reference
AD_LNEG Left justified with negative voltage reference
AD_RON Right justified. 6 Most Significant bits are read as '0'.
AD_RPOS Right justified with positive voltage reference
AD_RNEG Right justified with negative voltage reference

Var - is a variable that the conversion result will be returned in.

BRANCH index, [Label1,...LabelN]
Go to the Label specified by index.

Index - is a variable or constant that specifies the address to branch to.

Label1,...LabelN - are labels that specify where to branch to.

BUTTON pin, downstate, delay, rate, bytevariable, targetstate, address
Monitors a switch that may be active-low or active-high.

Pin - is a variable/constant that specifies the I/O pin to use. This pin will be made an input.

Downstate - is a variable/constant (0 or 1) that specifies which logical state occurs when the button is pressed.

Delay - is a variable/constant (0–255) that specifies how long the button must be pressed before auto-repeat starts. The delay is measured in cycles of the Button routine. Delay has two special settings: 0 and 255. If Delay is 0, Button performs no debounce or auto-repeat. If Delay is 255, Button performs debounce, but no auto-repeat.

Rate - is a variable/constant (0–255) that specifies the number of cycles between auto repeats. The rate is expressed in cycles of the Button routine.

Bytevariable - is the workspace for delay and rate.

Targetstate - is a variable/constant (0 or 1) that specifies which state the button should be in for a branch to occur. (0=not pressed, 1=pressed)

Address - is a label that specifies where to branch if the button is in the target state.

CLEAR
Clear user RAM.

COUNT pin, period, variable
Count the number of cycles (0-1-0) on the specified pin during period number of milliseconds and store that number in variable (Minimum of 4us pulse width).

Pin - is a variable/constant (0–15) that specifies the I/O pin to use.
This pin will be placed into input mode by writing a 0 to the corresponding bit of the DIRS register.

Period - is a variable/constant (1 to 4294967296) specifying the time in milliseconds during which to count.

Variable - is a variable (usually a word) in which the count will be stored.

DATA {@address,} value, {@address,} value
Data command will pre-load the internal EEPROM on the Atom during programming.

Address - is optional and specifies the first address to start at. Multiple address values can be used after each value is written. The Data command will start writing at the first location of the EEPROM by default. Address must be a number.

Value - are constant values that will be written to the Atom's on board EEPROM. These values must be constants since the values are placed during programming.

DEBUG [{Options} item, {{Options} item}]
Sends values of specified variables or constants to the debug watch window.

Options - are DEC, HEX, BIN or REAL. These modifiers will convert Item to DEC = Decimal, HEX = Hexadecimal, BIN = Binary digits or REAL = Value.

Item - can be a constant or variable. There is no limit to the amount of items used other than program memory.

DEBUGIN [(Options) item]
Receives byte values from the IDE DEBUG window and stores them in a specified variable on the Atom.

Options - are DEC, HEX, BIN or REAL. These modifiers will convert Item to DEC = Decimal, HEX = Hexadecimal, BIN = Binary digits or REAL = Value.
Item - can only be a byte variable. It stores the received byte value in the specified variable.

DO - WHILE
Repeat a group of commands while expression is true. True means the result of the expression is not equal to 0.

Expression - is any combination of variables, constants, math and logic operators.

DTMFOUT pin,{ontime,offtime,}[,tone...]
Generate dual-tone, multifrequency tones for DTMF (i.e., telephone "touch" tones).

Pin - is a variable/constant that specifies the I/O pin to use. This pin will be set to an output during tone generation. After tone generation is complete, the pin is left as an input.

Ontime - is an optional entry; a variable or constant (0 to 65535) specifying duration of the tone in milliseconds. If ontime is not specified, DTMFout defaults to 200 ms on.

Offtime - is an optional entry; a variable or constant (0 to 65535) specifying the length of silent pause after a tone (or between tones, if multiple tones are specified). If offtime is not specified, DTMFout defaults to 50 ms off.

Tone - is a variable or constant specifying the DTMF tone to send.
Tones 0 through 11 correspond to the standard layout of the telephone keypad, while 12 through 15 are the fourth-column tones used by phone test equipment and in ham-radio applications.

0 to 9 Digits 0 - 9
10 = *
11 = #
12—15 Fourth column tones A through D

DTMFOUT2 pin1 \ pin2,{ontime,offtime,}[,Tone...]
Uses two pins to generate dual-tone, producing a cleaner signal (i.e., telephone "touch" tones).

Pin1 \ Pin2 - is a variable/constant that specifies the I/O pins to use. These pins will be set as outputs, during tone generation. After tone generation is complete, the pins are set to inputs.

Ontime - is an optional entry; a variable or constant (0 to 65535) specifying a duration of the tone in milliseconds. If ontime is not specified, DTMFout defaults to 200 ms on.

Offtime - is an optional entry; a variable or constant (0 to 65535) specifying the length of silent pause after a tone (or between tones, if multiple tones are specified). If offtime is not specified, DTMFout defaults to 50 ms off.

Tone - is a variable or constant specifying the DTMF tone to send. Tones 0 through 11 correspond to the standard layout of the telephone keypad, while 12 through 15 are the fourth-column tones used by phone test equipment and in ham-radio applications.

0 to 9 Digits 0 - 9
10 = *
11 = #
12—15 Fourth column tones A through D

ENABLEVIDEO fontlib

This is actually a compiler directive, rather than a command. It loads the font table and sets the Atom to generate video. This requires some additional hardware to work but it essentially creates a video signal that can display characters on a TV screen.

fontlib is the filename of a predefined font bitmap library. This file should reside in your program directory on the PC, or else a complete path may be given. Sample font libraries are available from Basic Micro and example programs for games such as Pong can be downloaded from their website.

EXCEPTION label

If multiple exit points are needed from a subroutine, all but the last should use the EXCEPTION command. EXCEPTION differs from RETURN as follows:

RETURN Retrieves the saved address from the stack, clears the address from the stack, and sets program execution to the line following the GOSUB command.

EXCEPTION Clears the return address from the stack, and resumes program execution at the label given.

Label - is the label at which program execution should continue

END

This commands marks the end of the program. It can be inserted at the end of the main loop of code so subroutines and other commands can be listed after.

FOR variable = startVal to endVal {STEP stepVal} ... NEXT

Create a repeating loop that executes the program lines between FOR and NEXT, increment or decrement the variable according to stepVal, until the value of the variable passes the endVal.

Variable - is a bit, nibble, byte, word or long variable used as a counter.

StartVal - is a variable or constant that specifies the initial value of the variable.

EndVal - is a variable or constant that specifies the end value of the variable. When the value of the variable passes endVal execution stops and the program goes to the instruction after Next.

StepVal - is an optional variable or constant by which the variable increases or decreases with each trip through the FOR/ NEXT loop. Negative values for StepVal will decrement and positive values will increment.

For counter = 20 to 1 step -1 ; this will decrement -1
For counter = 20 to 1 ; this will over flow to 0 and count to 1
For counter = 1 to 20 step 1 ; this will increment +1
For counter = 1 to 20 ; this will increment +1 (assumed)

FREQOUT pin, duration, freq1{,freq2}

Generates one or two tones for a specified duration.

Pin - is a variable/constant that specifies the I/O pin to use. This pin will be put into output mode during generation of tones and left in that state after the instruction is completed.

Duration - is a variable/constant specifying the length in milliseconds (1 to 65535) of the tone(s).

Freq1 - is a variable/constant specifying frequency in hertz (Hz, 0 to 32767) of the first tone.

Freq2 - is a variable/constant specifying frequency (0 to 32767 Hz) of the optional second tone

GETWATCHDOG variable
Retrieves the last calculated Watchdog timeout value.

Variable - is a bit, nibble, byte, word or long variable which stores the timeout count.

Notes
This command retrieves a value established by the TIMEWATCHDOG command. The nominal period of the Watchdog timer is 18 ms. Since the timer is based on a simple R/C oscillator, and is not derived from the system clock, GETWATCHDOG gives you the ability to calculate an accurate watchdog timeout (using the SLEEP command) and to compensate for temperature variations.

GOSUB Label
Store the label after GOSUB, then go to the point in the program specified by Label. It must be used with the RETURN command.

Label - specifies the section of the program to jump to.

GOTO Label
Go to the point in the program specified by Label. (Label is a label that specifies where to go.)

Label - specifies the section of the program to jump to.

HIGH pin
Makes the specified pin an output and sets it to high (+5 volts is High).

Pin - is a variable/constant that specifies the I/O pin to use.

I2CIN DataPin, ClockPin,{ErrLabel,}Control,{Address,} [{mods}Var,...VarN]

Receives data from an I2C device (EEPROM, External A/D Converter)

DataPin - is a variable or constant that specifies the I/O pin to use for SDA.

ClockPin - is a variable or constant that specifies the pin that the MASTER I2C device will use to clock the bus signal. (SCL).

ErrLabel - is a label that the program will jump to if the I2CIN com mand fails (i.e.: device is not connected).

Control - is a variable or constant that specifies the I2C device's control byte. The control byte consists of two parts The first four bits are the device type (EEproms use %1010). The next three bits are the device ID. If the address lines of the serial EEPROM (i.e. : A0, A1, A2) are grounded then the next three bits of the control byte must be zero.(ie: %1010000). The last bit is a flag used to deter mine the addressing format, 1 =16bit addressing, 0 = 8bit addressing.

Address - is an optional variable or constant that specifies the starting address to read from.

Mods - are command modifiers, which can be used to modify the variable directly.

Var - is a variable or constant that specifies the data being sent from the current bus master.

VarN - is a list of variables and/or constants that specifies the data being sent from the current bus master.

I2COUT DataPin, ClockPin,{ErrLabel,}Control,{Address,} [{mods} Var,...VarN]

Sends data to an I2C device (EEPROM, External A/D Converter).

DataPin - is a variable or constant that specifies the I/O pin to use. (SDA)

ClockPin - is a variable or constant that specifies the pin that the MASTER I2C device will use to clock the bus signals. (SCL)

ErrLabel - is a label that the program will jump to if the I2CIN command fails, timeout occurs or a device is not connected.

Control - is a variable or constant that specifies the I2C device's control byte. The control byte consist of two parts The first four bits are the device type (EEproms use %1010). The next three bits are the device ID. If the address lines of the serial EEPROM (i.e. A0, A1, A2) are grounded then the next three bits of the control byte must be zero.(ie: %1010000). The last bit is a flag used to determine the addressing format,
1 =16bit addressing, 0 = 8bit addressing.

Address - is an optional variable or constant that specifies the starting address to write the data to.

Mods - are command modifiers, which can be used to modify the variable directly.

Var - is a variable or constant that specifies the data being sent from the current bus master.

VarN - is a list of variables and/or constants that specifies the data being sent from the current bus master. This allows for multiple variables to be written to the I2C device by automatically incrementing the last given address.

IF *Compare* THEN {Gosub} *Label*
Compare, if true(not 0) jump to label or:
IF *Compare* THEN (If condition is true then jump to label)
Statements... (If not true goto elseif)
ELSEIF *Compare* (If true execute statements)
*Statement*s... (If not true goto else)
Else (If nothing is true then execute statements)
Statements...
Endif (terminate after else statements ran)
The IF...THEN...ELSEIF...ELSE...ENDIF evaluates one or more conditions and, if true, jumps to a label. If false then skip next function.

Condition - is a statement, such as "x = 7" that can be evaluated as true or false.

Gosub - is optional. Choosing GOSUB allows you to return to the next line of your program after running a subroutine. The default is to jump to a label.

Label - is a label that specifies where to go in the event that the condition is true.

INPUT pin
Makes the specified pin an input

Pin - is a variable or constant that specifies the I/O pin to use.

LCDINIT regsel\clk {\RdWrPin}, nib

Initializes an LCD display. This command is used before using theLCDREAD or LCDWRITE commands but asa my project examples show, you can also setup the LCD with the LCDWRITE command.

regsel - is a constant or variable specifying the Atom I/O pin connected to the LCD's R/S line.

clk - is a constant or variable specifying the Atom I/O pin connected to the LCD's E (Enable) line.

LCDREAD RegSel\Clk\RdWrPin, Nib, Address, [{mods} Var]
Reads the RAM on a LCD module using the Hitachi 44780 controller or equivalent.

RegSel - can be a constant or variable specifying the pin for the R/S line of the LCD.

Clk - can be a constant or variable specifying the pin for the E (Enable) line of the LCD.

RdWrPin - can be a constant or variable specifying the pin for the R/W (Read / Write) line of the LCD.

Nib - can be a constant or variable specifying the pins (Total of Four) for the data lines of the LCD. The LCD data port is arranged in 8 bits. Only 4 are required 4 to 7. So if OUTA is selected then LCD bit 4 is connected to pin 0 on the Atom and LCD bit 5 is connected to pin 1 on the Atom and so on.

Address - can be a constant or variable that specifies the address location of RAM you are trying to read. Address from 0 to 127 return the current character in the display memory. Address 128 and above return Character RAM values.

Mods - are command modifiers, which can be used to modify the variable directly.

Var - is the variable where the value returned will be stored.

LCDWRITE RegSel \ Clk {\RdWrPin}, Nib, [{mods} Exp]
Sends Text to an LCD module using an Hitachi 44780 controller or equiva-lent.

RegSel - is optional for LCDWRITE and can be a constant or variable specifying the pin for the R/S line of the LCD.

Clk - can be a constant or variable specifying the pin for the E (Enable) line of the LCD.

RdWrPin - can be a constant or variable specifying the pin for the R/W (Read / Write) line of the LCD.

Nib - can be a constant or variable specifying the pins (Total of Four) for the data lines of the LCD. The LCD data port is arranged in 8 bits. Only 4 are required 4 to 7. So if OUTA is selected then LCD bit 4 is connected to pin 0 on the Atom and LCD bit 5 is connected to pin 1 on the Atom and so on.

Mods - are command modifiers, which can be used to modify the variable directly.

Exp - can be a constant or variable that is the data to be written.

LcdWrite Comand Table
There are several control commands that can be used with LCDWRITE such as CLEAR and HOME. Each additional control command used with LCDWRITE must be separated with a "," (comma) inside of the

brackets "[...]". Below is a chart of all the available control commands for use with LCDWRITE.

Command Name	Description
INITLCD1	Initialize LCD display
INITLCD2	Initialize LCD display
CLEAR	Clear Display
HOME	Return Home
INCCUR	Auto Increment Cursor(default)
INCSCR	Auto Increment Display
DECCUR	Auto Decrement Cursor
DECSCR	Auto Decrement Display
OFF	Display,Cursor, and Blink off
SCR	Display on,†Cursor and Blink off
SCRBLK	Display and Blink on, Cursor off
SCRCUR	Display and Cursor on, Blink off
SCRCURBLK	Display, Cursor, and Blink on
CURLEFT	Move Cursor left
CURRIGHT	Move cursor right
SCRLEFT	Move Display left
SCRRIGHT	Move Display right
ONELINE	Set display for 1 line LCDs
TWOLINE	Set display for 2 line LCDs
CGRAM + address	Set CGRAM address for R/W
SCRRAM + address	Set Display RAM address for R/W

LET Var = Value
Assign a value to a variable

Var - is the labeled variable

Value - can be a constant or another variable or the result of an expression which can be any (legal) combination of math operators.

LOOKDOWN
value,{comparisonOp,}[value0,value1,...valueN],resultVariable
Compare a value to a list of values according to the relationship specified by the comparison operator. Store the index number of the first value that makes the comparison true into resultVariable. If no value in the list makes the comparison true, resultVariable is unaffected

Value - is a variable or constant to be compared to the values in the list.

ComparisonOp - is optional and maybe one of the following:
= equal < less than
<> not equal>= greater than or equal to
> greater than <= less than or equal to
If no comparison operator is specified, Then MBasic defaults to equal (=).

Value0, value1... make up a list of values (constants or variables) up to 16 bits in size.

ResultVariable is a variable in which the index number will be stored if a true comparison is found.

LOOKUP index, [value0, value1,...valueN], resultVariable

Look up the value specified by the index and store it in a variable. If the index exceeds the highest index value of the items in the list, variable is unaffected. A maximum of 256 values can be included in the list.

Index - is the item number (constant or variable) of the value to be retrieved from the list of values.

Value0, value1... make up a list of values (constants or variables) up to 16 bits in size.

ResultVariable - is a variable in which the retrieved value will be stored.

LOW pin

Make the specified pin output low.

Pin - is a variable or constant that specifies the I/O pin to use.

NAP period

Enter sleep mode for a short period. Power consumption is reduced to about 50 μA assuming no loads are being driven. All pins are output low.

Period - is a variable or constant that determines the duration of the reduced power nap. The duration is $(2\wedge period) * 18$ ms. (Read as "2 raised to the power of 'period', times 18 ms.") Period can range from 0 to 7, resulting in the following nap lengths:

0 - 18ms
1 - 36ms
2 - 72ms
3 - 144ms
4 - 288ms
5 - 576ms
6 - 1152ms (1.152 seconds)
7 - 2304ms (2.304 seconds)

OUTPUT pin
This command makes the specified pin an output.

Pin - is a variable or constant that specifies the I/O pin to use.

OWIN Pin,Mode,{NCLabel,} [{Mods} Var]
Protocol used to communicate to 1-wire devices.

Pin - is a variable or constant that specifies the I/O pin to use for the One wire command.

Mode - is a variable/constant/expression indicating the mode of data transfer. Mode controls placement of reset pulses, detection of presence pulses, byte / bit input and normal / high speed. The proper value for Mode will depend on the 1-wire device used. Consult the device data sheet to determine the correct Mode. See chart below for reference:

Mode Setting
0 - No Reset, Byte mode, Low speed
1 - Reset before data, Byte mode, Low speed
2 - Reset after data, Byte mode, Low speed
3 - Reset before and after data, Byte mode, Low speed
4 - No Reset, Bit mode, Low speed
5 - Reset before data, Bit mode, Low speed

NCLabel - is a label the program can jump to if a connection failure occurs with the OWIN command (ie. No chip present).

Pin - is a variable or constant that specifies the I/O pin to use for the One wire command.

Mods - are command modifiers which can be used to modify the variable directly.

Var - is the variable or variable array where the value(s) returned will be stored.

OWOUT Pin,Mode,{NCLabel,} [{Mods} Exp]
Protocol used to communicate to 1-wire devices.

Pin - is a variable or constant that specifies the I/O pin used for the One wire command.

Mode - is a variable/constant/expression indicating the mode of data transfer. Mode controls placement of reset pulses, detection of presence pulses, byte / bit input and normal / high speed. The proper value for Mode will depend on the 1-wire device used. Consult the device data sheet to determine the correct Mode.
Reference chart below:

Mode Setting
0 - No Reset, Byte mode, Low speed
1 - Reset before data, Byte mode, Low speed
2 - Reset after data, Byte mode, Low speed
3 - Reset before and after data, Byte mode, Low speed
4 - No Reset, Bit mode, Low speed
5 - Reset before data, Bit mode, Low speed

NCLabel - is a label the program can jump to if a connection failure occurs with the OWOUT command (ie. No chip present).

Mods - are command modifiers which can be used to modify the variable directly.

Exp - is a variable, variable array, constant or expression containing the data to be sent.

PAUSE milliseconds

Pause the program (do nothing) for the specified number of milliseconds.

Milliseconds - is a variable or constant specifying the length of the pause in msec. milliseconds may be up to a 32 bit number.

PAUSECLK cycles

Pause the program (do nothing) for the specified number clock cycles divided by 4.

Cycles - is a variable or constant specifying the length of the pause. Pauses may be up to a 32 Bit number.

PAUSEUS microseconds

Pause the program (do nothing) for the specified number of micro seconds.

Microseconds - is a variable or constant specifying the length of the pause in μsec. Microseconds may be up to 65535 μs.

PEEK address, variable
POKE address, expression

Read/Write specified RAM location.

Address - is a variable or constant which denotes a memory location.

Variable - is a variable name and is where the results will be stored.

Expression - is any combination of variables, constants and math operations.

PULSIN pin, state, {TimeoutLabel,TimeoutMultiple,} Var

Measure the width of a pulse.

Pin - is a variable or constant that specifies the I/O pin to use. This pin will be placed into input mode during pulse measurement and left in that state after the instruction finishes.

State - is a variable or constant (0 or 1) that specifies whether the pulse to be measured begins with a 0-to-1 transition (1) or a 1-to-0 transition (0).

TimeoutLabel - is an optional label that specifies where to go if a time out occurs. The default time out value is 65,535 microseconds.

TimeoutMultiple - is a variable or constant that specifies the amount of time to wait before timing out. The time out value is multiplied by 65,535 microseconds which is the default value. If a value of 10 is used the command would wait 655,350 microseconds.

Var - is a variable in which the pulse duration will be stored.

PULSOUT pin, time
Output a pulse.

Pin - is a variable or constant that specifies the I/O pin to use.
This pin will be placed into output mode immediately before the pulse and left in that state after the instruction finishes.

Time - is a variable or constant (0-65535) that specifies the duration of the pulse in μs.

PUSH variable
PUSH Stores a 32 bit value on the stack.

POP, variable
POP Retrieves a 32 bit value from the stack.

Important: PUSH must always be matched by a subsequent POP instruction before any other stack-oriented commands (e.g. GOSUB, RETURN, EXCEPTION) are used.

103

variable - may be of any type. For *push* the value will be padded with high order zeros to fill 32 bits if necessary. For *pop* the high order bits will be truncated if necessary to fit the variable. Variable types should be matched for predictability. While it's possible, for example, to PUSH a long variable, and subsequently POP a word or byte variable, it's less confusing to stick to matched types. If you POP a word or byte variable only the low order bits will be stored.

PUSHW variable
PUSHW Stores a 16 bit value on the stack.

POPW variable
POPW Retrieves a 16 bit value from the stack.

PUSHW (push word) and POPW (pop word) are similar to PUSH and POP except that they deal with 16 bit, rather than 32 bit values.

Important: PUSHW must always be matched by a subsequent POPW instruction before any other stack-oriented commands (e.g. GOSUB, RETURN, EXCEPTION) are used.
In theory, it's possible to PUSHW a 16 bit address, then execute a RETURN to jump to that address, but it is not recommended since the GOTO command is much less confusing.

variable - may be of any type. For *push* the value will be padded with high order zeros to fill 16 bits if necessary. Longer variables will be truncated (high order bits lost) to fit 16 bits. For *pop* the high order bits will be truncated if necessary to fit the variable.

PWM pin, duty, cycles
Convert a digital value to analog output via pulse-width modulation.

Pin - is a variable or constant that specifies the I/O pin to use. This pin will be placed into output mode during pulse generation then switched to input mode when the instruction finishes.

Duty - is a variable or constant (0-255) that specifies the analog output level (0 to +5V).

Cycles - is a variable or constant (0-65535) specifying an approximate number of milliseconds of PWM output.

Variable = RANDOM expression (Seed Value)

Generate a pseudo random number. Random is a Math function.

Variable - is a variable that will store the results of the random command. This can be a Byte or Word sized variable depending on the expected results.

Expression - is any combination of variables, constants, math and logic operators.

RCTIME pin, state, {TimeoutLabel,TimeoutMultiple,}, resultVariable

Count time while pin remains in state—usually to measure the charge or discharge time of a resistor and capacitor circuit. (RC)

Pin - is a variable or constant that specifies the I/O pin to use. This pin will be placed into input mode and left in that statewhen the instruction finishes.

State - is a variable or constant (1 or 0) that will end the RCTIME period.

TimeoutLabel - is an optional label that specifies where to go if a time out occurs. The default time out value is 65,535 microseconds.

TimeoutMultiple - is a variable or constant that specifies the amount of time to wait before timing out. The time out value is multiplied by 65,535 microseconds which is the default value. If a value of 10 is used the command would wait 655,350 microseconds.

ResultVariable - is a variable in which the time measurement (0 to 65535 in µs units) will be stored.

READ location, variable

Read EEPROM location and store value in variable.

Location - is a variable or constant that specifies the EEPROM address to read from.

Variable - holds the byte value read from the EEPROM.

READDM address,[{mod}var,...,{mod}varN]
Read EEPROM location and store value in variable or variables.

Address - is a value pointing to a location in the eeprom from which to start.

Mod - is any appropriate input or output modifier.

Var - is any variable type.

REPEAT......UNTIL expression
Repeat a series of commands until the expression is true (value of expression > 0)

Expression - is any combination of variables, constants, math and logic operators.

REVERSE pin
Reverse the data direction of the specified pin.

Pin - is a variable or constant that specifies the I/O pin to use. This pin will be placed into the opposite of its current input/ output (I/O) mode.

Serdetect pin,mode,var
Detect incoming baud rate. Used for auto detecting baud rates.
Pin - is a variable or constant that specifies the I/O pin that will be used to receive the sync character.

Mode - is the settings for Bits 13,14 and 15. Bit 13 ($2000 hex) is a flag that controls the number of data bits and parity (0=8 bits and no parity, 1=7 bits and even parity). Bit 14 ($4000 hex) controls polarity (0=noninverted, 1=inverted). Bit 15 ($8000 hex) is not used by SERIN. Constants from the below table can be used for the Mode:

IMODE = Inverted

NMODE = Non Inverted
IEMODE = Inverted, Even Parity
NEMODE = Non Inverted, Even Parity
IOMODE = Inverted, Open Drain
NOMODE = Non Inverted, Open Drain
IEOMODE = Inverted, Even Parity, Open Drain
NEOMODE = Non Inverted, Even Parity, Open Drain

Var - is a word sized variable that will hold the calculated baudmode value which can be used by serin and serout.

SERIN rpin{\fpin},baudmode,{plabel,}{timeout,tlabel,}[inputData]
Receive asynchronous (e.g., RS-232) data.

Rpin - is a variable or constant that specifies the I/O pin through which the serial data will be received.

Fpin - is an optional variable or constant that specifies the I/O pin to be used for flow control. This pin will switch to output mode and remain in that state after the end of the instruction. (Caution Fpin requires the 'slash')

Baud mode - is a 16-bit variable or constant that specifies serial timing and configuration. The lower 13 bits are interpreted as the bit period. Bit 13 ($2000 hex) is a flag that controls the number of data bits and parity (0=8 bits and no parity, 1=7 bits and even parity). Bit 14 ($4000 hex) controls polarity (0=noninverted, 1=inverted). Bit 15 ($8000 hex) is not used by SERIN.

Plabel - is an optional label indicating where the program should go in the event of a parity error. This argument may only be provided if baud mode indicates 7 bits, and even parity, otherwise the label is ignored.

Timeout - is an optional variable/constant (0–65535) that tells SERIN how long, in milliseconds, to wait for incoming data. If data does not arrive in time, the program will jump to the address specified by Tlabel.

Tlabel - is an optional label which must be provided along with Timeout, indicating where the program should go in the event that data does not arrive within the period specified by Timeout.

InputData - is a list of variables and modifiers that tells SERIN what to do with incoming data. SERIN can store data in a variable or array; interpret numeric text (decimal, binary, or hex), and store the corresponding value in a variable; wait for a fixed or variable sequence of bytes; or ignore a specified number of bytes. These actions can be combined in any order in the inputData list.

SEROUT tpin\fpin,baudmode,{timeout,tlabel,}[outputData]

Transmit asynchronous (e.g., RS-232) data.

Tpin - is a variable or constant that specifies the I/O pin through which the serial data will be sent.

Baudmode - is a 16-bit variable or constant that specifies serial timing and configuration. The lower 13 bits are interpreted as the bit period. Bit 13 ($2000 hex) is a flag that controls the number of data bits and parity (0=8 bits and no parity, 1=7 bits and even parity). Bit 14 ($4000 hex) controls the bit polarity (0=noninverted, 1=inverted). Bit 15 ($8000 hex) determines whether the pin is driven to both states (0/1) or to one state and open in the other (0=both driven, 1=open).

Pace - is an optional variable or constant (0–65535) that tells SEROUT how long in milliseconds it should pause between transmitting bytes.

OutputData - is a list of variables, constants and modifiers that tells SEROUT how to format outgoing data. SEROUT can transmit individual or repeating bytes; convert values into decimal, hex or binary text representations; or transmit strings of bytes from variable arrays.

Fpin - is an optional variable or constant that specifies the I/O pin to be used for flow control (byte-by-byte handshaking). This pin will switch to input mode and remain in that state after the instruction is completed. (Caution Fpin requires the 'slash')

Timeout - is an optional variable/constant (0–65535) used in conjunction with fpin flow control. Timeout tells Serout how long in milliseconds to wait for fpin permission to send. If permission does not arrive in time, the program will continue at tlabel.

Tlabel - is an optional label used with Fpin flow control and timeout. Tlabel indicates where the program should go in the event that permission to transmit data is not granted within the period specified by the Timeout command.

SERVO pin, rotation{, repeat}

This command can control a model airplane style servo.

Pin is the pin controlling the servo.

Rotation - is a variable / constant that specifies the position you want the servo to rotate. A value from -1200 to + 1200 is used with 0 being center. Value -1200 rotates a servo to the farthest position in one direction and +1200 being the farthest rotation in the opposite direction. The maximum +1200 and minimum -1200 will vary based on the servo being used. Take caution not to exceed these values.

Repeat - (optional) Specifies the number of internal cycles the command runs(defaults to 30).

SHIFTIN Dpin,Cpin,Mode,[result{\bits}{,result{\bits}...}]

Shift data in from a synchronous-serial device.

Dpin - is a variable or constant that specifies the I/O pin that will be connected to the synchronous-serial device's data output. This pin's I/O direction will be changed to input and will remain in that state after the instruction is completed.

Cpin - is a variable or constant that specifies the I/O pin that will be connected to the synchronous-serial device's clock input. This pin's I/O direction will be changed to output.

Mode - is a value (0 - 7) or one of 8 predefined symbols that tells SHIFTIN the order in which data bits are to be arranged and the relationship of clock pulses to valid data. Here are the symbols, values, and their meanings:

MSBPRE or 0 - Data msb-first; sample bits before clock pulse
LSBPRE or 1 - Data lsb-first; sample bits before clock pulse
MSBPOST or 2 - Data msb-first; sample bits after clock pulse
LSBPOST or 3 - Data lsb-first; sample bits after clock pulse
FASTMSBPRE or 4 - Data msb-first; sample bits before clock pulse
FASTLSBPRE or 5 - Data lsb-first; sample bits before clock pulse
FASTMSBPOST or 6 - Data msb-first; sample bits after clock pulse

FASTLSBPOST or 7 - Data lsb-first; sample bits after clock pulse

(MSB is most-significant bit; the highest or left most bit of a nibble, byte, or word. LSB is the least-significant bit; the lowest or right most bit of a nibble, byte, or word.)

(Fast mode runs SHIFTIN at the fastest possible rate vs the normal mode of 100kbps)

Result - is a bit, nibble, byte, or word variable in which incoming data bits will be stored.
Bits - is an optional entry specifying how many bits (1—16) are to be input by SHIFTIN. If no bits entry is given, SHIFTIN defaults to 16 bits.

SHIFTOUT Dpin,Cpin,Mode,[data{\bits}{,data{\bits}...}]
Shift data out to a synchronous-serial device.

Dpin - is a variable or constant that specifies the I/O pin that will be connected to the synchronous-serial device's data input. This pin's I/O direction will be changed to output and will remain in that state after the instruction is completed.

Cpin - is a variable or constant that specifies the I/O pin that will be connected to the synchronous-serial device's clock input. This pin's I/O direction will be changed to output and will remain in that state after the instruction is completed.

Mode - is a value (0 or 1) or a predefined symbol that tells SHIFTOUT the order in which data bits are to be arranged. Here are the symbols, values, and their meanings:
Symbol Value Meaning:
MSBPRE or 0 - Data msb-first; sample bits before clock pulse
LSBPRE or 1 - Data lsb-first; sample bits before clock pulse
MSBPOST or 2 - Data msb-first; sample bits after clock pulse
LSBPOST or 3 - Data lsb-first; sample bits after clock pulse
FASTMSBPRE or 4 - Data msb-first; sample bits before clock pulse
FASTLSBPRE or 5 - Data lsb-first; sample bits before clock pulse
FASTMSBPOST or 6 - Data msb-first; sample bits after clock pulse
FASTLSBPOST or 7 - Data lsb-first; sample bits after clock pulse

(MSB is most-significant bit; the highest or left most bit of a nibble, byte, or word. LSB is the least-significant bit; the lowest or right most bit of a nibble, byte, or word.)

(Fast mode runs SHIFTIN at the fastest possible rate vs the normal mode of 100kbps)

Backwards Compatibility:
LSBFIRST 1 Data shifted out lsb-first.
MSBFIRST0 Data shifted out msb-first.

Data - is a variable or constant containing the data to be sent.

Bits - is an optional entry specifying how many bits (1—16) are to be output. If no bits entry is given, SHIFTOUT defaults to 16 bits.

SLEEP seconds
Put the Atom into low-power sleep mode for a specified number of seconds.

Seconds - is a variable or constant (1-65535) that specifies the duration of sleep in seconds.

Sound pin,[duration1\note1,...durationN\noteN]
Generate specific note from one pin.

Pin - is a variable/constant that specifies the I/O pin to use. This pin will be set to an output during tone generation and left in that state after the instruction is completed.

Duration - is a variable/constant specifying the length in milliseconds (1 to 65535) of the tone(s).

Note - is a variable/constant specifying frequency in hertz (Hz, 0 to 32767) of the first tone.

Sound2 pin1\pin2,[duration1\note1\note2_1,...durationN\noteN\note2_N]
Generate specific notes one on each of the two defined pins.

Pin1 \ Pin2 - is a variable/constant that specifies the I/O pins to use. This pin will be set to an output during tone generation and left in that state after the instruction is completed. The two specified pins can be tied together as shown in the schematic

Duration - is a variable/constant specifying the length in milliseconds (1 to 65535) of the tone(s).

Note - is a variable/constant specifying frequency in hertz (Hz, 0 to 16000) of the tones.

Sound8 port,[duration1\notelist1{,duration2\notelist2,...durationN\notelistN}]
Generates up to 8 simultaneous tones on 8 specified outputs.

Port - is a variable/constant that specifies the I/O port to use (outl p0 – p7, outh p8 – p15).

Duration - is a variable or constant defining the duration of each set of 8 tones in milliseconds (1 – 65535). All 8 tones in a set have the same duration. A new duration is specified for each set of 8 tones in a sequence.

Notelist - is a set of up to 8 variables or constants defining frequencies to be generated, one per output pin. The notelist is of the form note1{\note2{\note3{\… note8}}}. If fewer than 8 notes are included, they will be output on the lowest numbered pins of the port and the undefined pins will remain "silent". Each *note* is a variable or constant that specifies the frequency of the note in Hz (1 – 16000).

SPMOTOR pin, delay, step
This command controls a unipolar stepper motor.

Pin - can be a variable or constant. Pin specifies the first pin out of 4 control pins required. If P0 was used, the control pins would then be P0, P1, P2, P3.

Delay - can be a variable or constant and is a value from 0 to 65365 in milliseconds. Delay controls the speed at which the stepper motor will rotate. The delay will also vary from stepper motor to stepper motor.

Step - can be a variable or constant and is the number of steps and the direction. The direction is determined by the value of Step. Positive values being clockwise and negative numbers being counter clockwise. The step value can be a range from -32682 to +32682

STOP
Stops program execution.

SWAP variable,variable
Swaps the contents of two variables.

Variable - is the value to be swapped

TIMEWATCHDOG

Resets the Watchdog Timer and waits for a reset. It calculates the time between watchdog resets. The first command after the reset must be a GETWATCHDOG which will record the time taken.

TOGGLE pin
Invert the state of a pin.

Pin - is a variable or constant that specifies the I/O pin to use.

While...Wend
While expression is true do the commands between the while – wend.

Expression - is any combination of variables, constants, math and logic operators

WRITE address,byte
Write a byte of data to the EEPROM.

Address - is a variable or constant specifying the EEPROM address to write to.

Byte - is a data byte to be written into EEPROM.

WRITEDM address,[{mod}expression,...,{mod}expressionN]
Write sequentially to EEPROM locations.

Address - is a value pointing to a location in the EEPROM from which to start.

Mod - is any appropriate input or output modifier

Var - is any variable type.

Expression - is any legitimate math expression.

XIN DataPin\ZeroPin,House,{TimeoutLabel,TimeoutCount,}[{Modifier}Var]
Receive X-10 data and store keycode in a variable.

DataPin - is a variable/constant that specifies the I/O pin to use. This pin will be set to an input. The DataPin should be pulled high with a 4.7K resistor.

ZeroPin - is a variable/constant that specifies the I/O pin to use. This pin will be set to an input. The ZeroPin should be pulled high with a 4.7K resistor. The ZeroPin is used to detect the zero crossing timings from the X-10 device.

House - is a variable/constant used to filter out data with multiple house settings. House is a comparison value to match if the incoming data matches a particular house code otherwise the XIN command will continue waiting. The constant labels for each house code are as follows:

X_A	X_I
X_B	X_J
X_C	X_K
X_D	X_L
X_E	X_M
X_F	X_N
X_G	X_O
X_H	X_P

TimeoutLabel - is an optional label used to specify a place to jump to if a time out occurs.

TimeoutCount - is an optional value used in conjunction with the TimeoutLabel to specify the amount of time that occurs before jumping to the TimeoutLabel. Timeouts are based on commands from the X-10 module. The value placed for TimeoutCount will wait *N* amount of commands received from the X-10 module before a time out will occur.

Var - is a variable/constant used to store the results (KeyCode) of the XIN statement. The incoming data is the keycode which is 5 bits so a byte size variable is needed.

XOUT DataPin\ZeroPin, House, [{Unit}, {Modifiers} Keycode]
Transmits X-10 House code and Keycode.

DataPin - is a variable/constant that specifies the I/O pin to use. This pin will be set to an input. The DataPin should be pulled high with a 4.7K resistor.

ZeroPin - is a variable/constant that specifies the I/O pin to use. This pin will be set to an input. The ZeroPin should be pulled high with a 4.7K resistor. The ZeroPin is used to detect the zero crossing timings from the X-10 device.

House - is a variable/constant that corresponds to the House Code set on the X-10 module A through P. The constant labels for each house code are as follows:

X_A	X_I
X_B	X_J
X_C	X_K
X_D	X_L
X_E	X_M
X_F	X_N
X_G	X_O
X_H	X_P

Unit - is an optional variable/constant that specifies the address of the unit, 1 to 16. Unit codes are also considered KeyCodes.

Unit Constant
X_1   % 00110
X_2   % 00111
X_3   % 00100
X_4   % 00101
X_5   % 01000
X_6   % 01001
X_7   % 01010
X_8   % 01011
X_9   % 01110
X_10   % 01111
X_11   % 01100
X_12   % 01101
X_13   % 00000
X_14   % 00001
X_15   % 00010
X_16   % 00011

KeyCode - is a variable/constant that specifies the unit code or function. Multiple KeyCodes can be used in the XOUT command. Only certain commands will work in combination such as DIM codes.

Unit Constant
X_Units_On   % 10000
X_Lights_On   % 11000
X_On   % 10100
X_Off   % 11100
X_Dim   % 10010
X_Bright   % 11010
X_Lights_Off   % 10110
X_Hail   % 10001
X_Status_On   % 11011
X_Status_Off   % 10111
X_Status_Request % 11111

Advanced Commands

These commands use the special hardware peripherals and features of the Atom Micro.

HSERIN {label,timeout,}[InputData]
HSEROUT [OutputData]

Receive and Send Asynchronous RS-232 data.

label,Timeout(optional) - is a label that will be jumped to if the hserin buffer has no data available in the buffer and the timeout period has expired. If a timeout is not defined the hserin command will wait indefinitely while the uart receive buffer is empty.

InputData - can be a list of variables and or modifiers that tell HSERIN / HSEROUT what to do when sending or receiving data. All the modifiers supported by SERIN and SEROUT are supported by HSERIN and HSEROUT.

SETHSERIAL baudrate

Sets the baud rate of the hardware serial port, initializes the serial buffers and enables the hardware serial port interrupt handler. This command must be executed before any of hserin, hserout or hserstat are used. *Note: When using the hardware serial system the interrupts for the hardware serial port are not available.*

Baudrate - is any of the following:
H300 H12000 H28800 H115200
H600 H14400 H31200 H250000
H1200 H19200 H33600 H312500
H2400 H21600 H36000 H625000
H4800 H24000 H38400 H1250000
H9600 H26400 H57600

HSERSTAT funct{,label}

This command lets you check the status and/or clear the hardware serial port buffers. Before using this command you must use the SETHSERIAL command (see page 100) to set the correct baud rate.

Funct - is a value from 0 to 6 that determines the function of the hserstat command according to the following list:

0 - Clear input buffer
1 - Clear output buffer
2 - Clear both buffers
3 - If input data is available go to *label*
4 - If input data is not available go to *label*
5 - If output data is being sent go to *label*
6 - If output data is not being sent go to *label*

Label - is an optional argument (use with values 3 – 6) that specifies the destination jump address.

Hpwm CCPx, Period, Duty

Generate pulse-width modulation use internal hardware PWM.

CCPx - is a variable or constant of 0 or 1 that specifies the PWM hardware to use. The Atom has two PWM's available. The first is on pin 9 which is selected by setting a value of 1. The second is on pin 10 and is selected by using a value of 0 for CCPx.

Period - is a variable or constant from 0 to 16383 that specifies the period of the pulse width in CLK cycles.

Duty - is a variable or constant from 0 to 16383 that specifies the duty cycle of the pulse width.

SETPULLUPS mode

Enables or disables internal 10k pullups on pins 0 to 7

Mode - is PU_OFF to disable pullups or PU_ON to enable pull-ups.

SETEXTINT mode

SetExtInt sets the external interrupt pin to input and sets the state that will cause an interrupt (EXTINT must be enabled)

Mode - is the setting that will trigger the actual interrupt. There are two choices available:
EXT_H2L = Will activate when pin is pulled low (from high)
EXT_L2H = Will activate when pin is pulled high (from low)

SETTMR0 mode

SETTMR0 sets Timer0 mode. Timer0 is an internal 8 bit timer module built into the Atom.

Mode - is the setting that will trigger the actual interrupt. There are several options available:
TMR0INT1 Internal timer with 1:1 timing
TMR0INT2 Internal timer with 1:2 timing
TMR0INT4 Internal timer with 1:4 timing
TMR0INT8 Internal timer with 1:8 timing
TMR0INT16 Internal timer with 1:16 timing
TMR0INT32 Internal timer with 1:32 timing
TMR0INT64 Internal timer with 1:64 timing
TMR0INT128 Internal timer with 1:128 timing
TMR0INT256 Internal timer with 1:256 timing
TMR0EXTL1 External counter(Low to High transition on AX3 pin)with 1:1 timing
TMR0EXTL2 External counter(Low to High transition on AX3 pin) with 1:2 timing
TMR0EXTL4 External counter(Low to High transition on AX3 pin) with 1:4 timing
TMR0EXTL8 External counter(Low to High transition on AX3 pin) with 1:8 timing
pin) with 1:16 timing
TMR0EXTL32 External counter(Low to High transition on AX3 pin) with 1:32 timing
TMR0EXTL64 External counter(Low to High transition on AX3 pin) with 1:64 timing
TMR0EXTL128 External counter(Low to High transition on AX3 pin) with 1:128 timing.
TMR0EXTL256 External counter(Low to High transition on AX3 pin) with 1:256 timing.
TMR0EXTH1 External counter(High to Low transition on AX3 pin) with 1:1 timing
TMR0EXTH2 External counter(High to Low transition on AX3 pin) with 1:2 timing
TMR0EXTH4 External counter(High to Low transition on AX3 pin) with 1:4 timing

TMR0EXTH8 External counter(High to Low transition on AX3 pin) with 1:8 timing
TMR0EXTH16 External counter(High to Low transition on AX3 pin) with 1:16 timing
TMR0EXTH32 External counter(High to Low transition on AX3 pin) with 1:32 timing
TMR0EXTH64 External counter(High to Low transition on AX3 pin) with 1:64 timing
TMR0EXTH128 External counter(High to Low transition on AX3 pin) with 1:128 timing
TMR0EXTH256 External counter(High to Low transition on AX3 pin) with 1:256 timing

SETTMR1 mode

SetTmr1 sets Timer1 mode. Timer1 is an internal 16 bit timer module built into the Atom.

Mode is the setting that will trigger the actual interrupt. There are several options available:
TMR1OFF Disables Timer1(saves power/default on powerup)
TMR1INT1 Internal timer with 1:1 timings
TMR1INT2 Internal timer with 1:2 timings
TMR1INT4 Internal timer with 1:4 timings
TMR1INT8 Internal timer with 1:8 timings
TMR1EXT1 External osc with 1:1 timings
TMR1EXT2 External osc with 1:2 timings
TMR1EXT4 External osc with 1:4 timings
TMR1EXT8 External osc with 1:8 timings
TMR1ASYNC1 External Asynchronous counter with 1:1 timings
TMR1ASYNC2 External Asynchronous counter with 1:2 timings
TMR1ASYNC4 External Asynchronous counter with 1:4 timings
TMR1ASYNC8 External Asynchronous counter with 1:8 timings

RESETTMR1 expr

Lets you reset Timer1 to a specified value (16 bits)
Expr - is a variable, constant or expression that determines the starting value for Timer1.

Notes

This command lets you pre-set Timer1 to a value greater than the default 0 so that it will time out in a shorter period. Normally the timer will overflow from 65535 to 0, generating an interrupt. If, for example, you set *expr* to 15535, Timer1 will overflow in 50000 clock ticks.

SETTMR2 mode, period

SetTmr2 sets Timer2 mode and reset period. Timer2 is an internal 8 bit timer with an 8 bit period module built into the Atom.

Mode - is the setting that will trigger the actual interrupt. There are several options available.

Period - is a reset point. Period will cause the timer reset when timer equals period. Period is a value of 0 to 255

Modes are:
TMR2OFF Disables Timer2 default on powerup
TMR2PRE1POST1 1:1 prescaler and 1:1 postscaler
TMR2PRE1POST2 1:1 prescaler and 1:2 postscaler
TMR2PRE1POST3 1:1 prescaler and 1:3 postscaler
TMR2PRE1POST4 1:1 prescaler and 1:4 postscaler
TMR2PRE1POST5 1:1 prescaler and 1:5 postscaler
TMR2PRE1POST6 1:1 prescaler and 1:6 postscaler
TMR2PRE1POST7 1:1 prescaler and 1:7 postscaler
TMR2PRE1POST8 1:1 prescaler and 1:8 postscaler
TMR2PRE1POST9 1:1 prescaler and 1:9 postscaler
TMR2PRE1POST10 1:1 prescaler and 1:10 postscaler
TMR2PRE1POST11 1:1 prescaler and 1:11 postscaler
TMR2PRE1POST12 1:1 prescaler and 1:12 postscaler
TMR2PRE1POST13 1:1 prescaler and 1:13 postscaler
TMR2PRE1POST14 1:1 prescaler and 1:14 postscaler
TMR2PRE1POST15 1:1 prescaler and 1:15 postscaler
TMR2PRE1POST16 1:1 prescaler and 1:16 postscaler
TMR2PRE4POST1 1:4 prescaler and 1:1 postscaler
TMR2PRE4POST2 1:4 prescaler and 1:2 postscaler
TMR2PRE4POST3 1:4 prescaler and 1:3 postscaler
TMR2PRE4POST4 1:4 prescaler and 1:4 postscaler
TMR2PRE4POST5 1:4 prescaler and 1:5 postscaler
TMR2PRE4POST6 1:4 prescaler and 1:6 postscaler
TMR2PRE4POST7 1:4 prescaler and 1:7 postscaler
TMR2PRE4POST8 1:4 prescaler and 1:8 postscaler
TMR2PRE4POST9 1:4 prescaler and 1:9 postscaler

TMR2PRE4POST10 1:4 prescaler and 1:10 postscaler
TMR2PRE4POST11 1:4 prescaler and 1:11 postscaler
TMR2PRE4POST12 1:4 prescaler and 1:12 postscaler
TMR2PRE4POST13 1:4 prescaler and 1:13 postscaler
TMR2PRE4POST14 1:4 prescaler and 1:14 postscaler
TMR2PRE4POST15 1:4 prescaler and 1:15 postscaler
TMR2PRE4POST16 1:4 prescaler and 1:16 postscaler
TMR2PRE16POST1 1:16 prescaler and 1:1 postscaler
TMR2PRE16POST2 1:16 prescaler and 1:2 postscaler
TMR2PRE16POST3 1:16 prescaler and 1:3 postscaler
TMR2PRE16POST4 1:16 prescaler and 1:4 postscaler
TMR2PRE16POST5 1:16 prescaler and 1:5 postscaler
TMR2PRE16POST6 1:16 prescaler and 1:6 postscaler
TMR2PRE16POST7 1:16 prescaler and 1:7 postscaler
TMR2PRE16POST8 1:16 prescaler and 1:8 postscaler
TMR2PRE16POST9 1:16 prescaler and 1:9 postscaler
TMR2PRE16POST10 1:16 prescaler and 1:10 postscaler
TMR2PRE16POST11 1:16 prescaler and 1:11 postscaler
TMR2PRE16POST12 1:16 prescaler and 1:12 postscaler
TMR2PRE16POST13 1:16 prescaler and 1:13 postscaler
TMR2PRE16POST14 1:16 prescaler and 1:14 postscaler
TMR2PRE16POST15 1:16 prescaler and 1:15 postscaler
TMR2PRE16POST16 1:16 prescaler and 1:16 postscaler

SETCAPTURE ccppin,mode

Sets up the capture hardware of the Atom.

Ccppin - specifies which module to use 0 for CCP1 on pin 10 or 1 for CCP2 on pin 9

Mode - is the setting that will trigger the actual interrupt. There are several options available.

CAPTUREOFF Disables Capture default on powerup
CAPTURE1H2L Captures Timer1 value on High to Low transition
CAPTURE1L2H Captures Timer1 value on Low to High transition
CAPTURE4L2H Captures Timer1 value on 4th Low to High transition
CAPTURE16L2H Captures Timer1 value on 16th Low to High transition

GETCAPTURE ccppin,var
Retrieves a previously captured timer value.

Ccppin - specifies which module to use 0 for CCP1 on pin 10 or 1 for CCP2 on pin 9
Var - is a word sized variable that the results of the 16 bit capture value will be returned to.

SETCOMPARE ccppin,mode,compare value
Sets up the compare mode in the Atom.

Ccppin - specifies which module to use 0 for CCP1 on pin 10 or 1 for CCP2 on pin 9

Compare Value - is the value that the comparison must match before the mode's action will occur.

Modes are:

COMPAREOFF	Disables Compare default on powerup
COMPARESETHIGH	Compare sets CCPx pin high on Timer1 comparison match
COMPARESETLOW match	Compare sets CCPx pin low on Timer1 comparison
COMPAREINT	Compare sets Interrupt(CCPxINT) and clears Timer1 on Timer1 comparison match
COMPARESPECIAL	Compare runs Special(Reset Timer1(CCP1) or Reset Timer1 and Activate(if A/D is enabled) A/D conversion(CCP2) on Timer1 comparison match

ENABLE intname
DISABLE intname
Enables or disables one or all interrupts. ENABLE must be used before interrupts will work. DISABLE prevents the specified interrupt from working. To use the External or CCP interrupts, SETEXTINT or SETCOMPARE must be used in addition to the ENABLE command to configure the Atom's hardware.

Intname - must be one of the interrupt names from the table below. Intname is optional, if it is omitted all interrupts will be enabled or disabled.

EXTINT - External (on I/O pin P0) INTF
RBINT - RB/OnChange (on I/O pins P4 – P7) RBIF
TMR0INT - Timer0 T0IF
TMR1INT - Timer1 TMR1IF
TMR2INT - Timer2 TMR2IF
ADINT - A/D conversion ADIF
RCINT - Receive RCIF
TXINT - Transmit TXIF
SSPINT - Sync Serial SSPIF
CCP1INT - Capture/Compare/PWM CCP1IF
CCP2INT - Capture/Compare/PWM CCP2IF
EEINT - EEPROM write complete EEIF
BCLINT - I2C bus collision BCLIF

SETEXTINT mode

Sets up the Atom's hardware to use P0 for external interrupt. This command must be used in addition to the ENABLE command before the External interrupt will work. (If SETEXTINT is not used, P0 will remain a conventional I/O pin and will not generate an interrupt.)

Mode - is one of the following:
EXT_H2L - interrupt on a high to low transition
EXT_L2H - interrupt on a low to high transition

ONINTERRUPT intname, label

This is a compile time function that establishes the label interrupt will jump to when it occurs. You must also enable the interrupt before it will work.

Label - is the label to which program execution will jump when this interrupt occurs. *Note: You must use the RESUME command (see below) to return to normal program execution after your interrupt has been processed.*

Intname - must be one of the interrupt names from the table below. Intname is optional, if it is omitted all interrupts will be enabled or disabled.

EXTINT - External (on I/O pin P0) INTF
RBINT - RB/OnChange (on I/O pins P4 – P7) RBIF
TMR0INT - Timer0 T0IF
TMR1INT - Timer1 TMR1IF
TMR2INT - Timer2 TMR2IF
ADINT - A/D conversion ADIF
RCINT - Receive RCIF
TXINT - Transmit TXIF
SSPINT - Sync Serial SSPIF
CCP1INT - Capture/Compare/PWM CCP1IF
CCP2INT - Capture/Compare/PWM CCP2IF
EEINT - EEPROM write complete EEIF
BCLINT - I2C bus collision BCLIF

ONPOR label
ONBOR label
ONMOR label

Establishes the place to jump to when these different resets occur. These commands allow your program to have different starting points depending on which type of reset has occurred. These commands are processed at compile time.

POR - Power On Reset Generates an interrupt when power is applied.

BOR - Brown Out Reset Generates an interrupt when voltage falls below 4.2 volts and then returns to normal.

MOR - Master Reset (MCLR) or Watchdog Timer Reset (WDT)
This reset generates an interrupt if the ATN or RES pin on the Atom module are pulled low, or the Watchdog Timer times out.

RESUME

Return from interrupt. This command is used to return to the point in your program where execution was interrupted. It should be used at the end of the interrupt processing code for ONINTERRUPT.

Chapter 4 – Getting Started

Although you can choose any Atom module to perform the experiments in this book, I chose to use my own Ultimate OEM module design. There are more reasons than "because I designed it", though I did design it based on what I wanted in an Atom module. I wanted to use a module that was as complete as possible and thus limit the amount of equipment the user had to buy. The real advantage to the Ultimate OEM is the ability to easily jump from one low cost breadboard to another leaving the rest of the circuit available for future re-use.

The picture below shows a complete Ultimate OEM starter package and this can be purchased from the sources at my website (www.elproducts.com) for typically less than $100 with the Atom manual on CD rather than printed as shown in the picture.

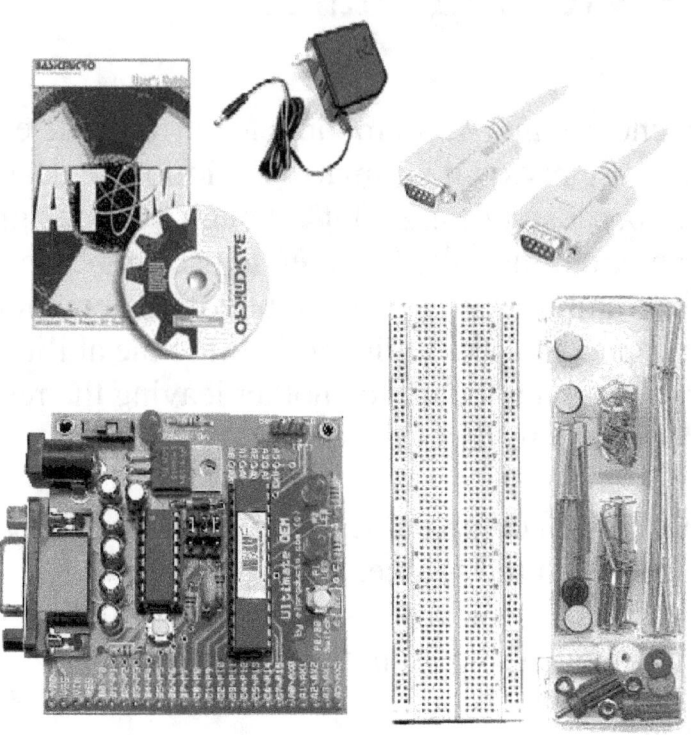

Some people prefer the Atom 24 pin modules because they have Basic Stamp boards lying around and that is fine for most experiments except when you need access to the A/D pins. In addition, the Atom 24 is surface mount design which is difficult to repair if a voltage regulator or PIC should fail on the board. The Ultimate has the Atom micro in a socket so you can easily replace it or even remove it and build it into a permanent design. New Atom micro interpreter chips are cheaper than buying a new 24 pin module.

An Atom 24 pin development board costs a lot more than a $10 breadboard like the one I use in the experiments of this book. I like to build up a circuit and then leave it assembled the same way you might write software and save it as a file on your hard drive for future use. The

Ultimate OEM makes that breadboard idea easy to work with by plugging into the breadboard as shown in the PICture below.

All the Atom I/O is brought out the header that plugs into the breadboard so connections are really easy to understand. The Ultimate also has the power adapter port, programming connector and on/off switch built in which are usually part of the development board for the Atom 24, 28 and 40 pin versions.

Some of the features built into the Ultimate OEM are shown below. The Ultimate is also upgradeable to a bootloader board for working with a native Basic Compiler or even programmable in assembly. This is beyond this book but it's nice to know you are not limited by the capability of the hardware.

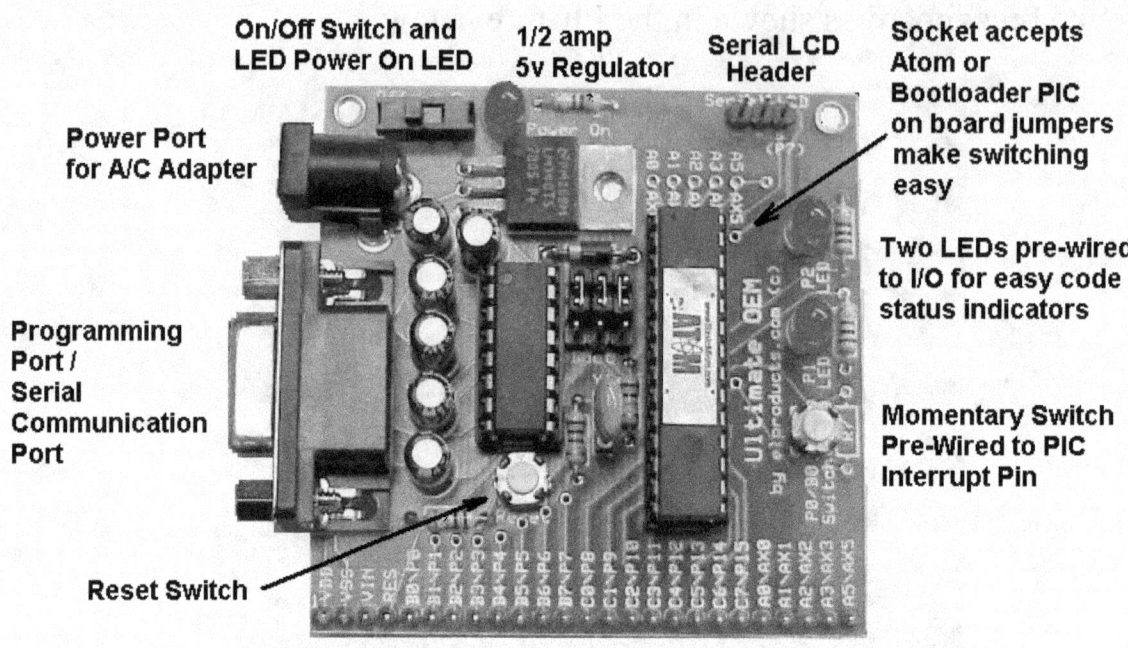

Power Port for A/C Adapter

On/Off Switch and LED Power On LED

1/2 amp 5v Regulator

Serial LCD Header

Socket accepts Atom or Bootloader PIC on board jumpers make switching easy

Programming Port / Serial Communication Port

Two LEDs pre-wired to I/O for easy code status indicators

Momentary Switch Pre-Wired to PIC Interrupt Pin

Reset Switch

21 I/O with 90 deg header for easy breadboard insertion

130

The schematic for the Ultimate OEM is shown below so you can understand how I built the module. The extra LEDs and momentary switch are there for convenience. I use one of the LEDs often for monitoring that a program is running properly. If the LED isn't flashing then the program is frozen somewhere.

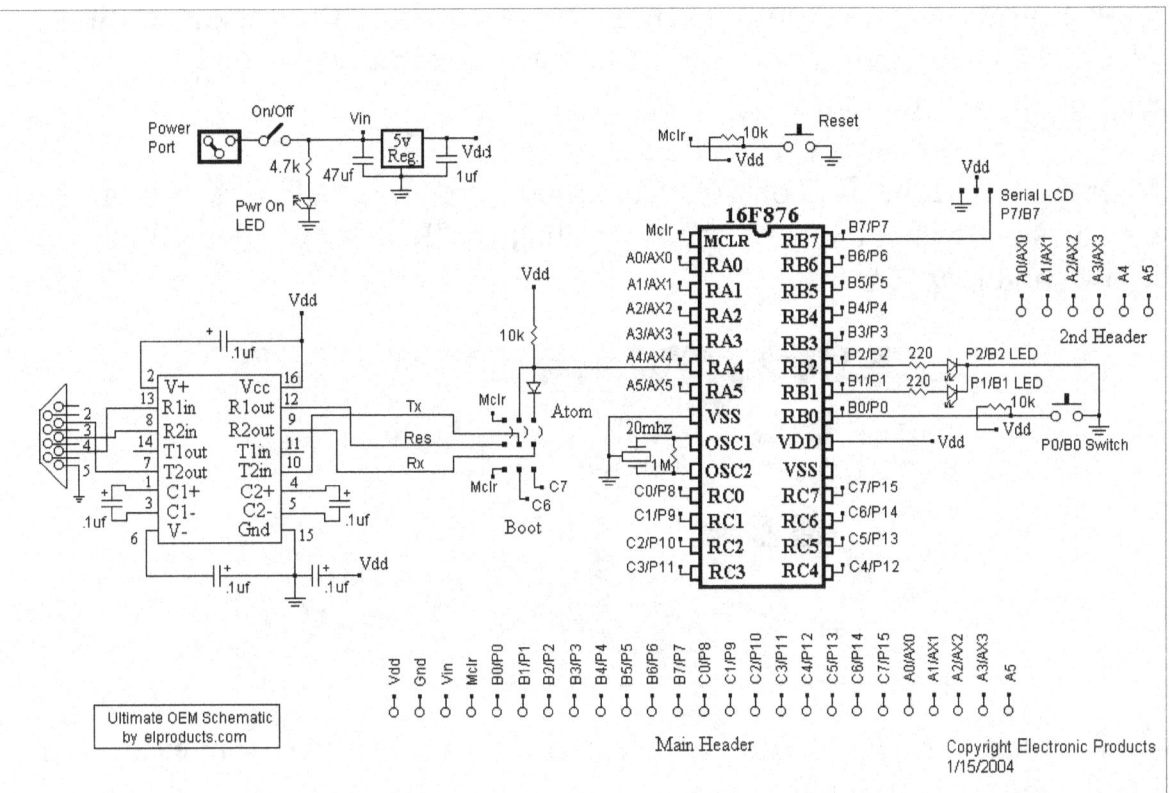

131

Installing the Atom programming software

First step is to copy all the necessary files into a directory on your PC hard drive. You can download the files from my website at www.elproducts.com/atombookfiles. If you purchase the packaged kit the included CD contains all the files. Insert the CD into your computer and copy those files to your PC.

When you have the files on your computer, click on the "Start" button in windows and when the menu pops up click on "Run". The dialog box below should appear.

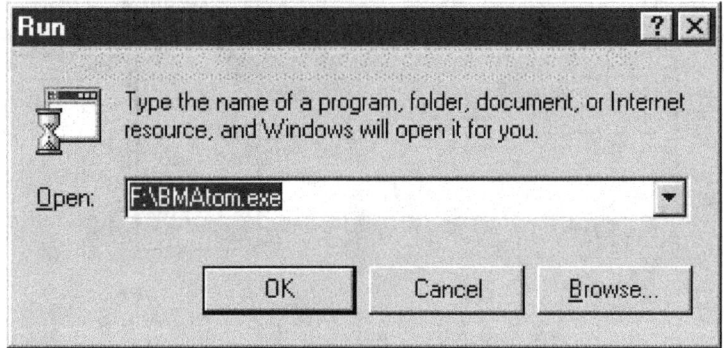

Click on the "Browse" button and search for the file "BMAtom.exe" in the list of files you copied. Find the file and then click on it.

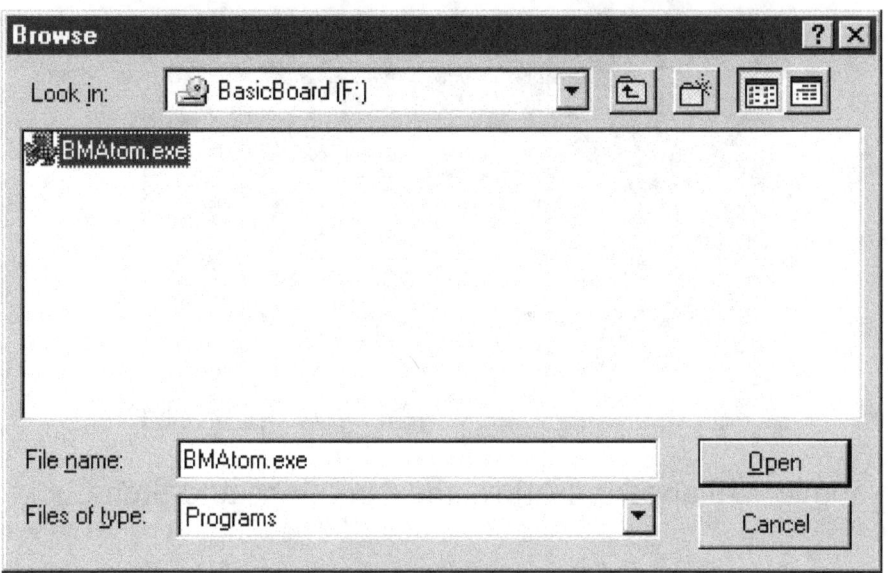

Then click on "Open". The "Run" screen should re-appear. Click on the "Run" button and the software should start installing itself and then bring up the screen below.

Click on the "Next" button.

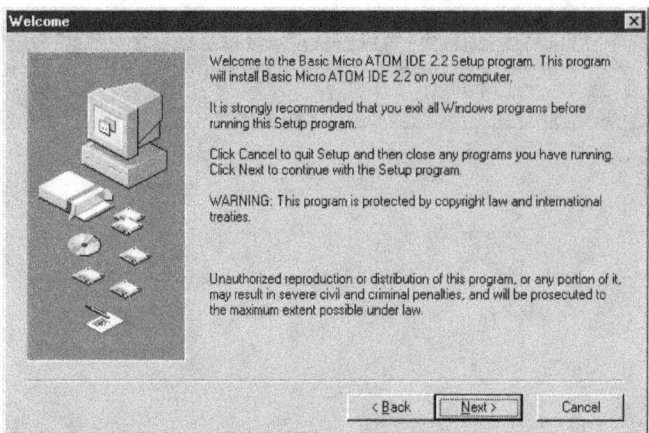

The screen above will appear. Follow the directions and then click on "Next".

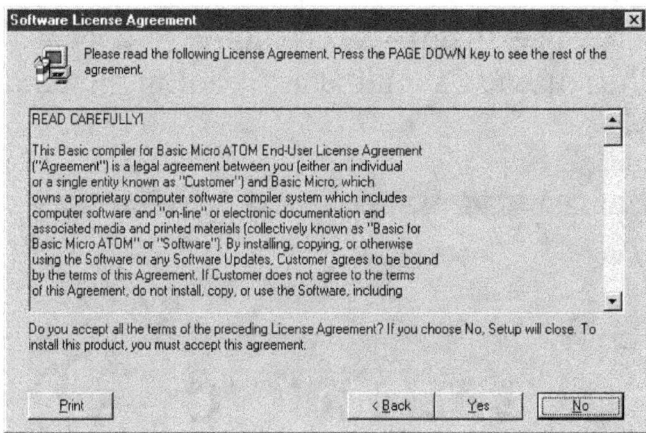

The next screen to appear is the License agreement. Read it over and if you agree, then click on "Yes".

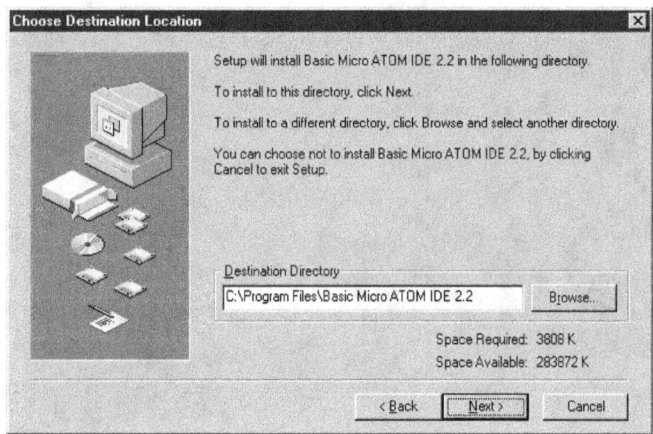

After you click on "Yes" the screen above will pop up to show you where the software will be installed on your hardrive.
Adjust the location if you want. Click on "Next" when you're ready.

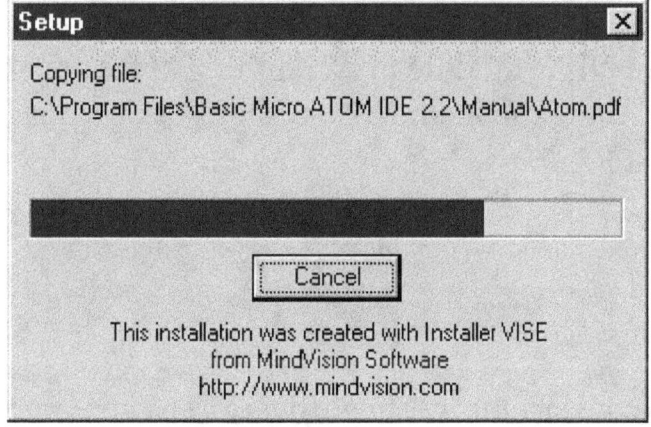

The software will begin to load and a solid bar will show you the progress. This should not take very long at all.

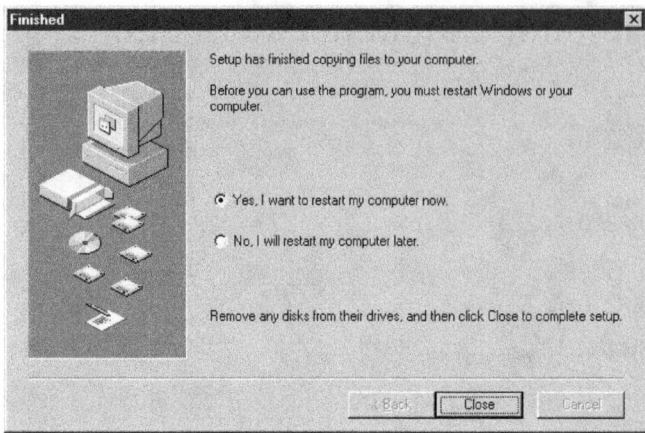

Finally the screen above will appear asking to restart your computer. Click on "Close" and your computer should reboot with the software installed.

Operation
The Atom/Ultimate OEM module programming software should now be listed in the main "Start" menu and also a shortcut on your PC's desktop screen.

Click on either one to start the Atom software. The screen below will appear.

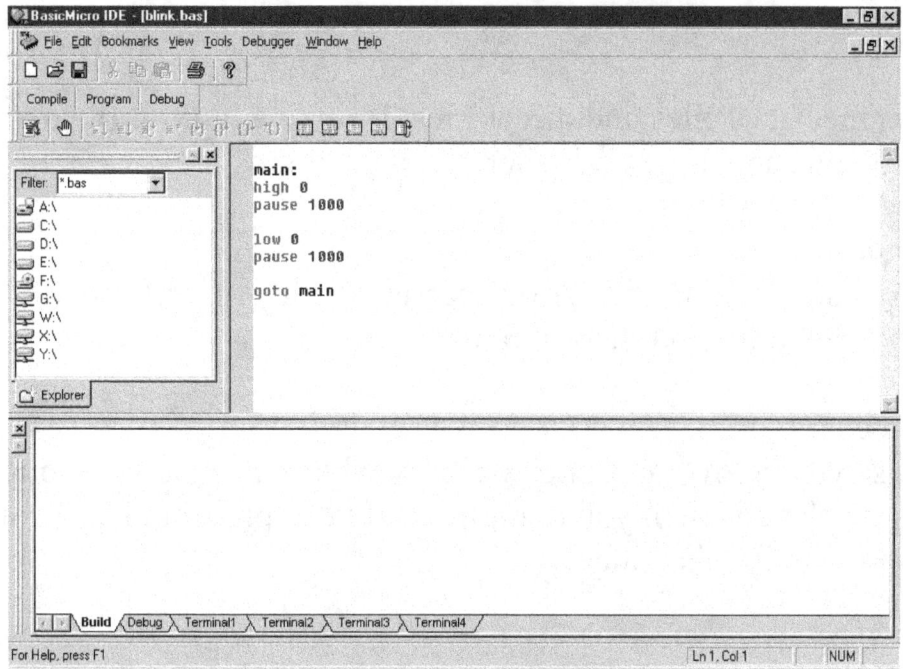

Tool Bar

This is the top of the screen where the standard windows menu is, with the new document icon, open folder icon, disk for saving icon, etc.

User Tools

Next is the "Compile", "Program", "Debug" buttons for compiling your Ultimate OEM module program.

Debugger Bar

The next line has several unique symbol icons. These are used for the debugger that you can use to watch your program run command by command. By placing the cursor over those symbols, you can see what each of them represents.

Workspace
This area is the list of files and directories. It's like a built in file manager window. So you can easily find where your programs are.

Editor Window
The main window next to the Workspace is where you will actually write and modify the Atom module programs.

Output Screen
This is the screen below the Editor window where the status of your program is displayed when you compile it. If your program has errors, they will show up in this window.

Status Bar
Finally at the very bottom is the Status Bar that spells out where the cursor is. It describes what is happening when you compile and generally, just tells you the status.

All these can be shut off, with the exception of the Editor Window, by clicking on the "View" selection at the top screen menu under the "Toolbars" sub menu item.

Special Note:
The sample programs on the CD or must be copied to your PC hard-drive before using them. The Atom software tries to save the file before downloading it to the Atom chip. If the file is loaded into the Atom software directly from the CD, an error will appear that the file cannot be saved. This is caused by the write protection on the CD.

Load 1st program

To load your first program you click on the folder symbol in the tool bar.

A window similar to the one below should appear. Change the directory to where you copied the sample files from the CD.

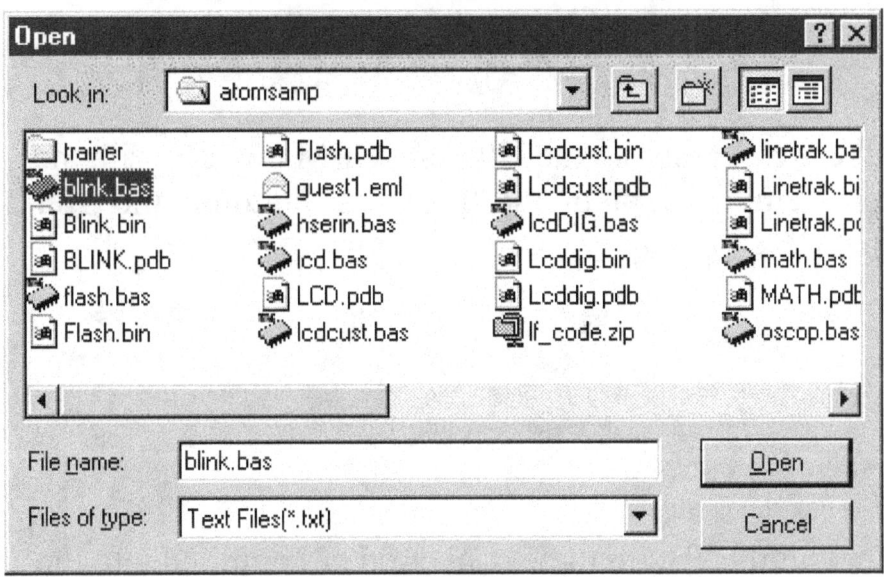

Find and highlight the blink.bas file and click on the "Open" button.

The Blink.bas file should now be in the Editor window. This program will flash LED1 on the Ultimate OEM module on for 1 second and then off for 1 second. This program is ready to be compiled and then programmed into the Ultimate OEM module. Before you take that next step though, let's check the system setup.

Click on the "Tools" menu item and look for "System Setup". Click on "System Setup" and the screen below should appear. Verify that the proper COM port is shown in the window that you connected the Ultimate OEM module serial cable to. Leave the debugger window alone. You only have to check this the first time you use the software.

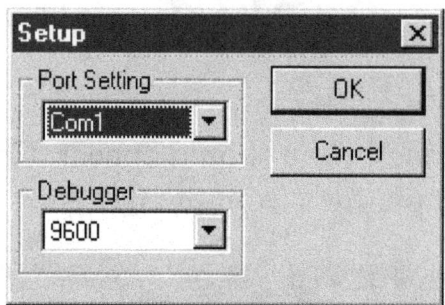

Now you're ready to program the Ultimate OEM module with the Blink.Bas program. Clicking on the "Program" button in the User Tools area does that.

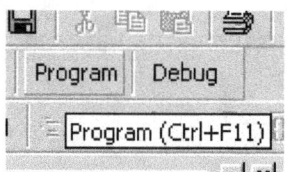

The program will then begin to be converted or compiled into the code required by the Atom micro. The output window will display what is happening and the first screen to show up will be the assembler window. If any errors are present, the output window will show them. If no errors are detected, the screen below will appear.

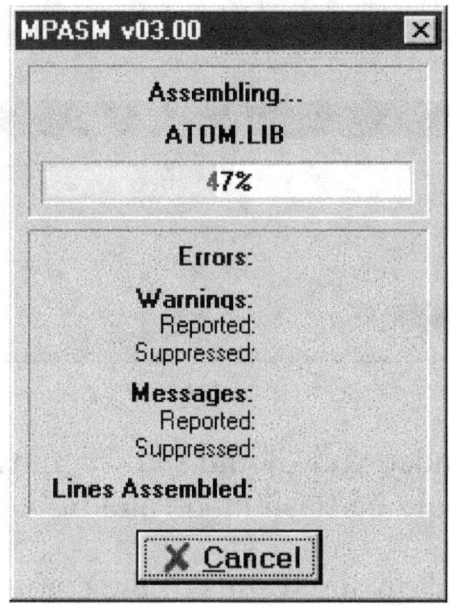

After the assembling screen, the software will begin to load the converted program into the Ultimate OEM module. That will happen in two steps. First the programming window will appear, and then the verification window. It will appear as the two screens below.

Writing to the Ultimate OEM module.

Verifying what was written to the Ultimate OEM module.

Finally the program is loaded and should start running on the Ultimate OEM module. LED1 should be flashing on and off continuously.

The output window will show the results of the compiling process similar to the PICture below.

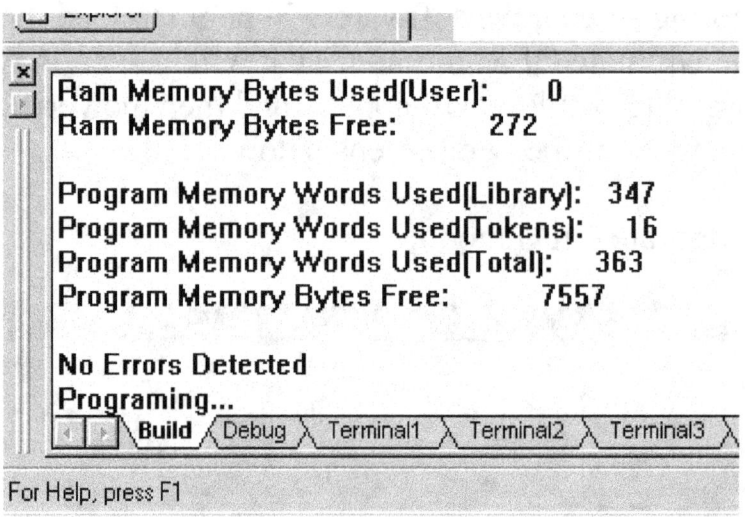

The screen has several sections of information. The first is the RAM Memory. RAM memory is where the program variables contents are stored. This screen display above indicates how many variable bytes the program used and how many are left. In this example above, 0 were used and 272 bytes are still available for future modifications. The results show zero used because the blink.bas program doesn't use any variables.

The amount of variable space used will also depend on what size your variables are. Byte variables will take up a solid byte, where word variables or floating point variables will take more space (two bytes and four bytes respectively).

The next section of information reported is the program memory. Unlike the RAM memory, program memory is reported in word size rather than byte size. Each command uses a different number of words and it's dependent on how much the command is doing. A HIGH command doesn't use many words of program memory space but a LCDWRITE command does several things beyond just controlling an I/O pin, therefore it uses several words of space.

The [Library] and [Tokens] values reported are not that useful for most programmers but the [Total] is. This value represents how much of the Atom micro's program memory you have used. The maximum amount of space is 7,920 words. This program used only 363 words and 7,557 are still available.

Although this program is small, 363 words may seem like a large amount of program space, and it is. The reason the value is so large is due to the internal setup of variables and constants along with command structure setup. These functions take a few hundred words of space at the beginning of every program. Once that setup is complete though, only the space required by the commands increase the program space. You will find that 7,920 words of total program space is a lot of program space and will allow very large programs to be written into the Ultimate OEM module.

Congratulations, If you got this far you have officially programmed the Ultimate OEM Atom micro.

In-Circuit Debugger (ICD)

One the great advantages to the Basic Atom software is the In-Circuit Debugger (ICD). With the ICD you can watch your program run in the Atom chip/module and find any bugs or programming mistakes as they happen. It's like making everything happen in slow motion so you can see where the error occurred. Once you use it you won't ever want to develop without it.

The ICD is completely controlled by software. No additional hardware is required. When you have a completed program, you would normally press the "Program" button to compile and download your program into the Atom chip/module. To use the ICD you simply press the "Debug" button instead. The difference between the "Program" button and the "Debug" button is hidden. They both compile and then program the Atom chip/module but the "Debug" adds another hidden step, it adds a block of code to your program that is used by the ICD.

When your program is running, the added ICD block of code sends variable values, internal register values and other details to the PC through the programming cable. The ICD, built into the programming screen IDE, will display that data in the way you require, by using the ICD debug tool bar. These functions allow you to run the program automatically, command by command, in slow motion or step through the program by using the PC mouse to advance the program command by command. This gives you total control of how the program advances.

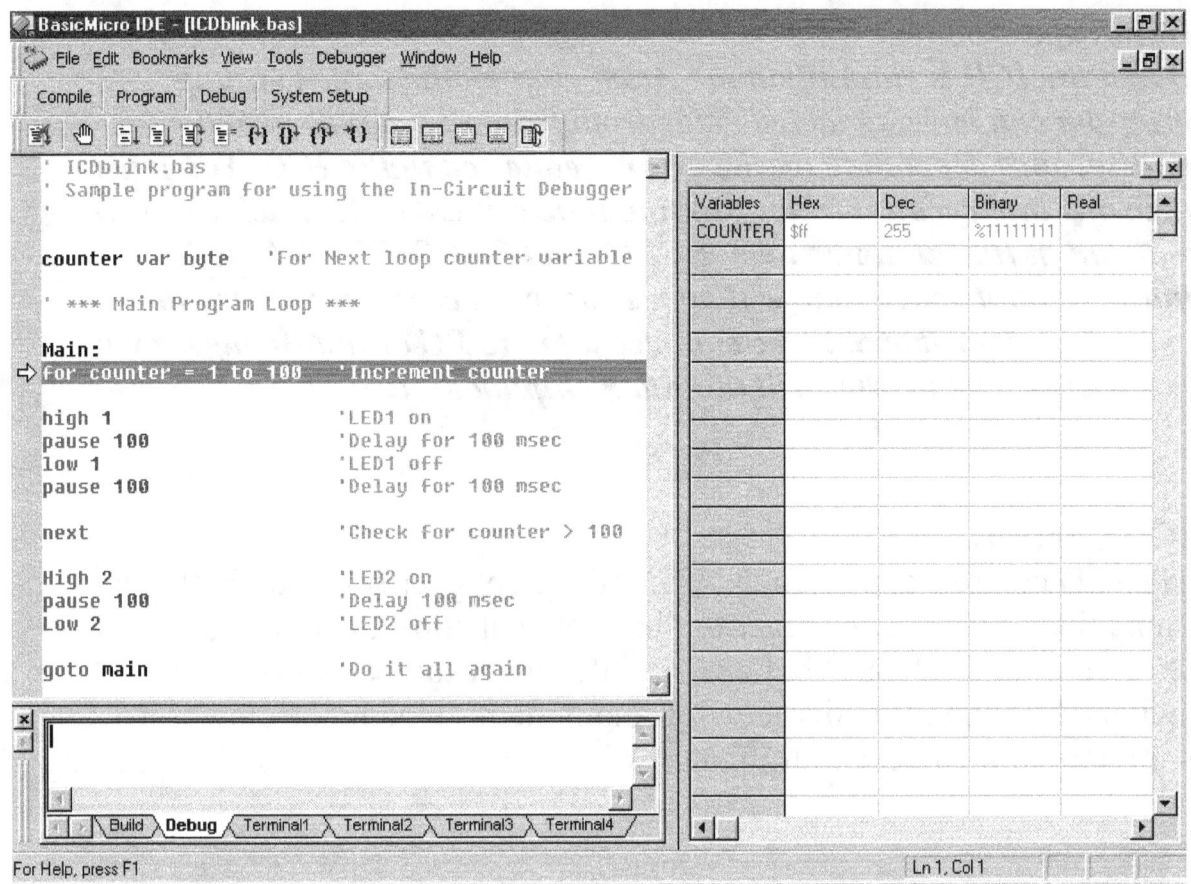

The ICD also allows you to view the state of each variable, each output state and even allows you to monitor the inner workings of the Atom chip/module (this requires knowledge of the Microchip PIC microcontroller which the Atom chip/modules Atom Chip is built from).The ICD screen shown above is a program running in DEBUG mode.

Important Note

When the ICD is running in the Atom chip/module, the running program can have an added delay from 0.5 milliseconds to 500 milliseconds depending on the action requested in the ICD. You must take this into account when running time critical code. Each command will run in full runtime mode (SERIN and SEROUT will function normally) but added time will appear between commands. Also the programming cable must be connected to the PC or the Debugger will not operate and neither will the Atom chip/module.

ICD Controls

The ICD controls and be found under the Debugger selection in the top menu line or via the debugger toolbar line that appears below the "Program" and "Debug" buttons. The debugger toolbar line can be switched on and off under the view main menu selection. Below is a summary of the ICD control features.

Connect/Disconnect
The Connect/Disconnect button is used to establish communication between the ICD and the Atom chip/module. When the "DEBUG" button is pressed and the program is downloaded, the ICD will automatically connect the ICD to the Atom chip/module. A green bar will highlight the first line of the program indicating the ICD is successfully connected. The ICON will change to Disconnect so you can disconnect the ICD from the Atom chip/module at any time.

Toggle Breakpoint

The Toggle Breakpoint button allows you to turn a breakpoint on or off at any point in the program. A breakpoint is a highlighted line that will stop execution of the program when it gets to that command line. This is handy if you want to see what variables and I/O pins look like when a specific command is encountered in the program without having to step through each command. To use it, just position the cursor to the line you want the program to stop at. If a break point is not set on that line, right click on your mouse and select "Toggle Breakpoint". This will highlight that command line in red and enable the breakpoint action. To turn it off, just right click on it again to turn it off.

Animate

This is a nice feature of the ICD. The Animate function will automatically step through your program command by command in slow motion. Each command being executed is highlighted in green. When the command is completed the next line is highlighted. This allows you to watch and verify the program is flowing where you expect to go. If the "Auto Update" feature is selected (described below) then variables and internal information will be updated after each command. To view those values though, you will have to pause the program as the animate mode runs too fast to read the data.

Run

This option allows you to run the program in the Atom chip/module at full speed (minus minor delay for the debugger block of code) without stopping to check for variables or other data. The green command line indicator will not step through each command.

Reset

Reset is used to start the program at the beginning. Any information stored in variables is not erased. This is a simple way to start at the beginning or to see how your program will react if a hardware reset were to occur.

Pause

The pause button will halt the program at the current command line. To resume execution, the RUN or ANIMATE button is pressed. The PAUSE button is handy to stop the RUN or ANIMATE mode so variables and other data can be viewed.

Step Into

This is the button you press to step through your program command by command line using your PC mouse.

Step Over

This button is a special step button that allows you to jump over a part of the program such as a gosub or for-next routine. Sometimes a gosub or for-next routine will take many clicks of the mouse to get through the routine using Step Into. This allows you to jump over it and move on to the command lines after them.

Step Out

This is another special step button that allows you to leave a gosub routine. It's handy for looking at part of a gosub routine and then lets you leave when you have seen enough. Clicking this will jump you to the command line after the end of the gosub routine.

Run To Cursor

Clicking on any command line in the program will produce a blinking cursor. If you then click on the "Run To Cursor" button, the program will execute in "RUN" mode until the cursor line is encountered. The program execution will stop at that command line.

Show Variables

This control button will toggle the Variables window open or closed. When it's selected a separate window will open and the variables defined in your program will automatically be listed. The values of those variables will be displayed in HEX, Decimal and Binary formats. (Make sure auto update is selected so these are updated after every command).

Show SFRs

SFR stands for Special Function Registers. These are special internal locations within the Atom chip that indicate how the internal program is controlling the Atom's Microchip PIC microcontroller. This is really a function for the advanced user but can be handy for understanding how the Atom Basic program controls the Microchip PIC microcontroller.

Show RAM
This feature shows all the Random Access Memory in the Atom chip/module, not just the variables. Again this is handy for the advanced user to see the inner workings of the Microchip PIC microcontroller.

Show Gosub Stack
This displays the Gosub Stack. The gosub stack is the list of location pointers within the Microchip PIC microcontroller that directs where to jump to when a gosub command is encountered. By monitoring this you can make sure multiple gosubs are not somehow getting lost. This is really an advanced user function.

Set Auto Update
This should always be selected. It tells the ICD to update the variable, RAM, SFRs and Stack after every command is executed. You should select this when the debugger is first connected but it can be turned on at anytime.

ICD Example

To demonstrate how to use the ICD we will use a different version of the blink.bas example used earlier that flashed an LED1. This program is very similar but uses a variable to count the number of flashes and changes both the state of the I/O pins and the counter variable. The program will flash LED1 on the Ultimate OEM 100 times and then light LED2 before looping back to do it all again. The variable "counter" stores the number of flashes so we can use the ICD variable window to watch the counter variable value change. The program is listed below.

```
' ICDblink.bas
' Sample program for using the In-Circuit Debugger
'

counter var byte      'For Next loop counter variable

' *** Main Program Loop ***

Main:
For counter = 1 to 100    'Increment counter

High 1                'LED1 on
Pause 100             'Delay for 100 msec
Low 1                 'LED1 off
Pause 100             'Delay for 100 msec

Next                  'Check for counter > 100

High 2                'LED2 on
Pause 100             'Delay for 100 msec
Low 2                 'LED2 off

Goto main             'Do it all again
```

Entering ICD Debug Mode

Load the program into the editor window. When you've completed that, make sure your Atom chip/module is connected to the programming cable and is powered up. Next press the "DEBUG" button. If everything compiles and programs properly, the window debug mode will appear with the first command line highlighted in green.

Step Into

Press the "auto update" button and also press the "show variables" button to open the variables display window. The variable "counter" should appear in the window. Now press the "step into" button to advance the program. Keep stepping through the program and watch the LED turn on then turn off several times. After you've seen the LED flash on and off several times, look at the "counter" variable in the variable window and see if the value of counter has changed to match the number of times the LED has flashed (it may be one larger depending on where you stopped the program).

Animate and Breakpoint

Now set the cursor to the command line with "next" in it and press the "Toggle Breakpoint" button. The "next" command line should turn red as seen in the PICture. Once that is completed, press the "Animate" button to make the program run. The program should step through the main loop and then stop at the "next" command line. Click on "animate" again and the program will stop at the breakpoint again. Watch the variable counter change with each stop at the breakpoint.

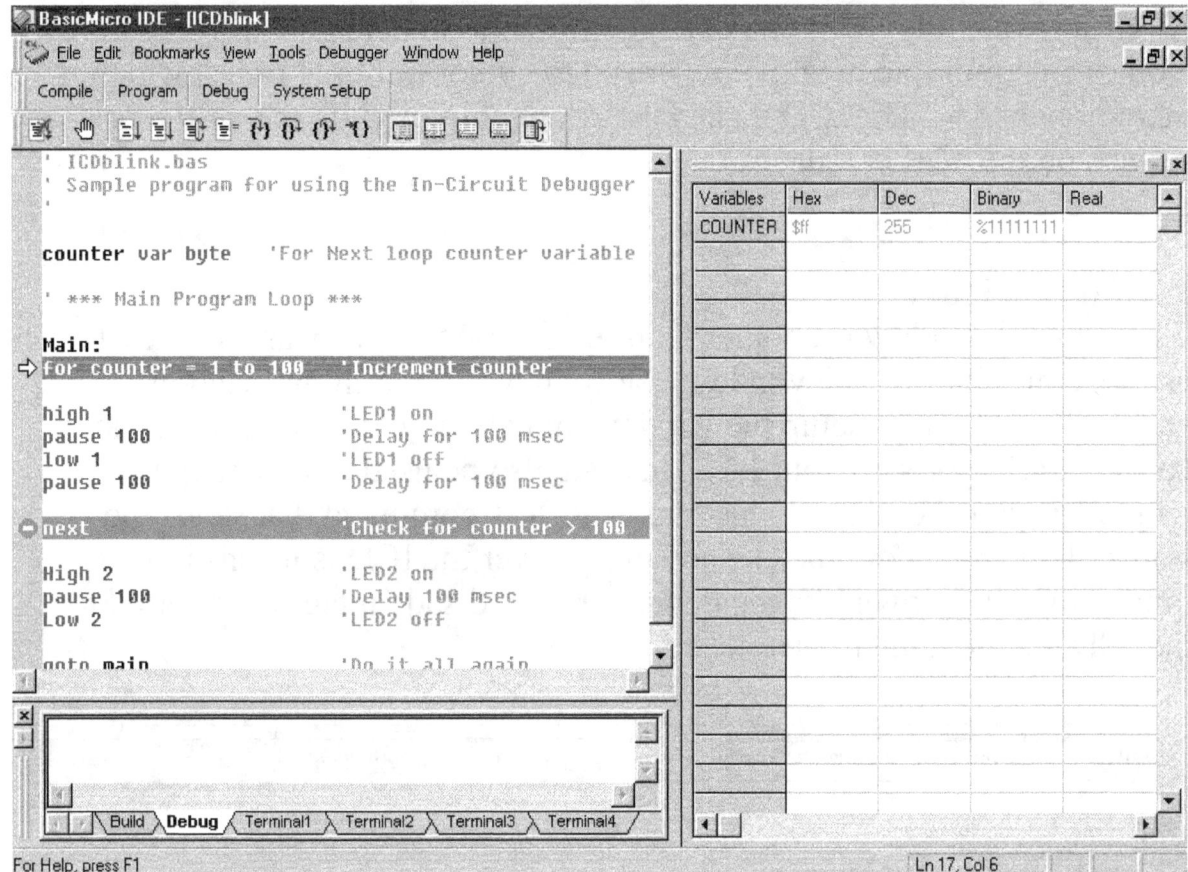

Reset and Step Over

Click on "Reset" to make the program jump back to the beginning. Now press "Step Over" and watch the program jump past the For-Next loop and stop at the "High 2" command line. This is how you bypass a lengthy section of code to see other sections run.

Summary

The example presented above is quite simple but hopefully you can see how handy the ICD is for debugging your program. As your program grows, the complexity also increases. Trying to find out why a program that runs, but isn't doing what you expect, is almost impossible to find without some help. The ICD is priceless and yet it's included free with the

153

Atom software. Play with the debugger often to understand all it's features. You can set multiple breakpoints or run to cursor multiple times, which is incredibly handy. The ICD makes the Atom chip/module one of the best set-ups on the market today.

Terminal Window

The terminal window is another feature of the Atom software. In fact it has several of these. The window below shows the terminal window. It has a selection bar to setup the window and allow you to connect to the various COM ports on your PC. This can also be used as a debugger window if you occasionally insert a SEROUT command that sends the status of a variable or pin. The advantage over the ICD is it runs in real time not slowed down like the debugger. The disadvantages compared to the ICD are too numerous to list.

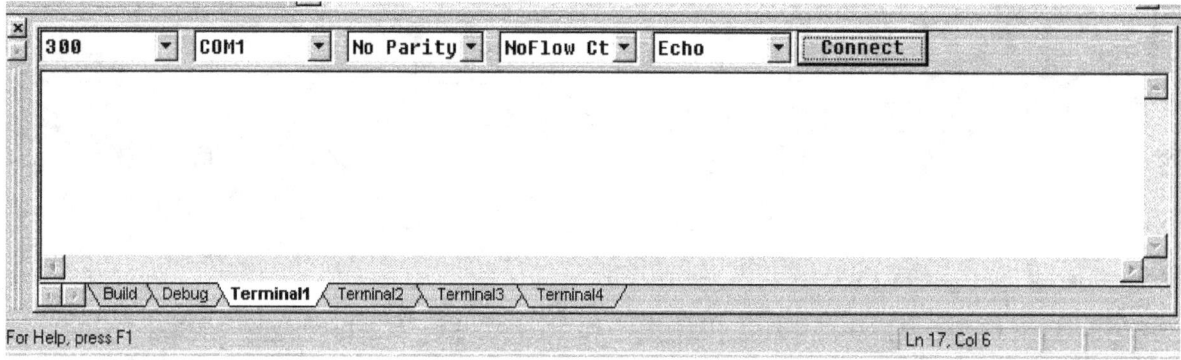

In Chapter 13, I use this window to communicate with the Ultimate OEM module. Since I assume most people know what a terminal window is I won't go into great depth. For those that don't just recognize that a terminal window is a way to display the ASCII data received by the serial port.

Terminal Control

The terminal window can be setup in various formats. The biggest options are the baud rate shown in the first pull-down box. Here you match this up to the baud rate your SEROUT command will use.

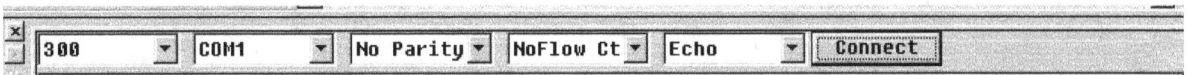

Next you can select the proper COM port and then setup the other communication parameters such as Parity or No Parity, flow control or no flow control and also if you want the things you type in the terminal window to echo back and display within the terminal window after you hit the enter key.

The last item is the "Connect/Disconnect" button. When you are ready to receive or send data to an Atom module, you would connect by clicking on the connect button. You won't be able to program the Atom if you use the programming port S_IN and S_OUT in your SEROUT command as the communication port until you disconnect.

You can have more than one terminal window running also and you select them from the tabs at the bottom of the terminal window as seen below.

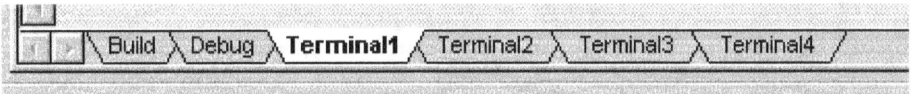

The terminal window is a handy tool and having it part of the Atom development window is very handy. You will use it often.

Chapter 5 – Flash an LED

Description

This is a very simple program but also a great starting point to make sure the software and the Ultimate OEM module are working together properly. This project is the same function as we used to setup the Ultimate OEM but this time we control an LED connected to the breadboard instead of the Ultimate OEM LED.

This project will flash an LED, connected to P0, on and off at a one second rate. The firmware chip does this by cycling the P0 pin between a high level (5-volts) and low level (ground). A one second delay is inserted between the commands that set the P0 pin high and low.

Project setup

The Ultimate OEM module has the I/O pins brought out to a 90 degree header that plugs easily into a breadboard. The connections of the LED and series 220 ohm resistor connect between the P0 pin and ground or Vss on the Ultimate OEM header. Also make sure the power adapter is plugged into the Ultimate OEM module and the wall socket.

The schematic below shows the LED properly wired to the P0 pin.

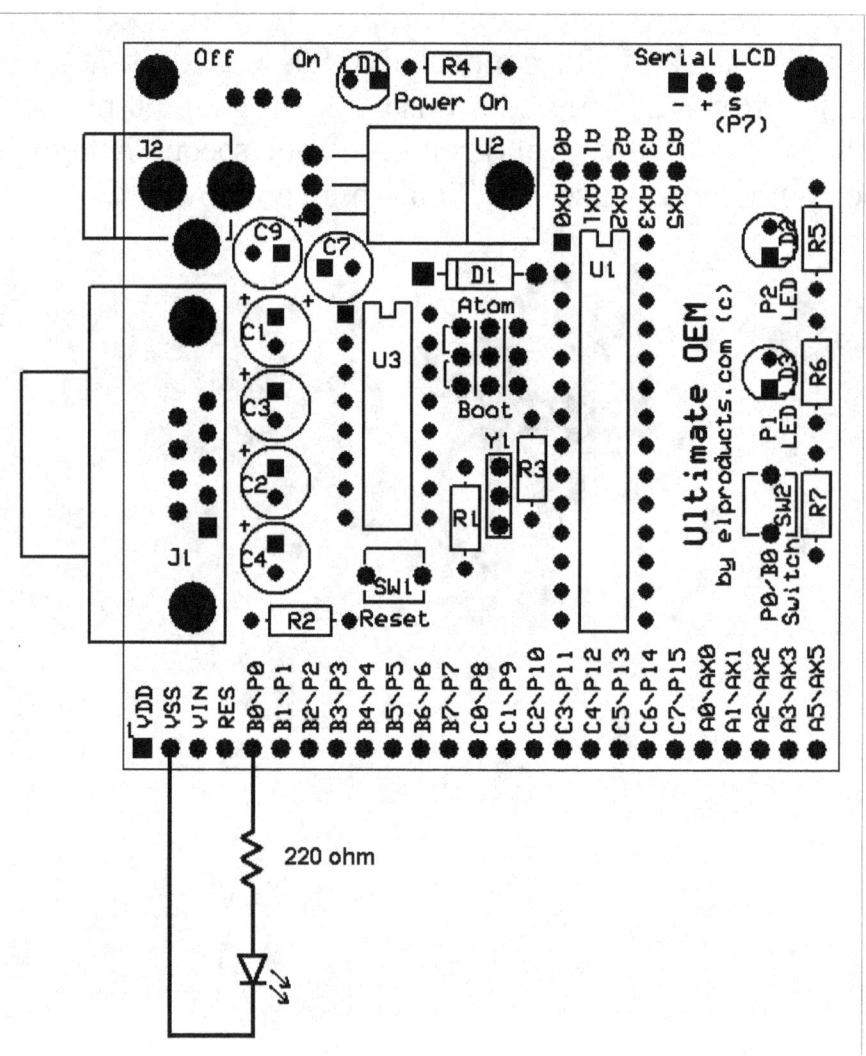

220 ohm

Software

The program listing below will perform the flash LED0 function. You can type the program in or you can download it off the included disk.

```
' *** Main program loop ***

main:                    ' main loop label
high P0                  ' Turn LED on
pause 1000               ' Delay for 1 second

low P0                   ' Turn LED off
pause 1000               ' Delay for 1 second

goto main   ' Loop back to main label and do it again

end                      ' End of the program
```

How it works

The software program controls the LED and flashes it at the one-second rate. To understand the software, lets step through the major sections of code.

An apostrophe in front of the description line tells the Atom compiler that the line is a comment and not a command. After the description the label "main" marks the beginning of the software.

```
' *** Main program loop ***

main:
```

This label "main" defines a location, within the program, where the main section of code begins. Under this label is where the LED is functioned. The HIGH and LOW command do most of the work. The HIGH command is used to output 5-volts to the LED. The LOW command is used to output ground or 0-volts to the LED. Both commands are followed by P0 which is the label for the Atom pin connected to the LED.

```
high P0     'Set LED on

low P0      'Set LED off
```

Because the HIGH and LOW commands are so quick, a delay between these commands is required to make the LED blink visible. The amount of time the LED is off or on is controlled by the PAUSE command.

The PAUSE command does nothing but delay. The PAUSE command includes a number that determines how many milliseconds (1/1000's of a second) to delay. In this case 1000 is used which represents 1000 milliseconds or 1 second.

```
pause 1000   ' Delay for 1 second
```

Finally the program goes into a loop so the whole process can be completed over and over again. Using the GOTO command generates the loop. The GOTO command is used to jump program control back to the top of the loop or to the "main" label.

```
goto main  ' Loop back to main label and do it again
```

To make sure the program does not get lost, the END command is added at the bottom. If this command is ever encountered, the program will stop running until the reset switch is pressed or power is removed and reconnected.

Next steps

A simple next step is to change the delay time from one second to something much faster or slower. Another option is to add another LED. Just repeat the group of HIGH, LOW and PAUSE commands to the main loop. Then modify the pin number of the second group to drive a different pin with a separate LED connected. You can alternate flashing the LEDs similar to a train track warning signal. This is a great modification to this simple project.

Chapter 6 – Scroll LEDs

Description

This project expands on the first project and drives eight LEDs using the P8 thru P15 Ultimate OEM pins, but not at the same time. This project will light one LED at a time in order to make the light scroll across the LEDs. Then the direction will reverse and the LEDs will scroll in the opposite direction. If we write the software to do this in a continuous loop, it will make the light scroll back and forth.

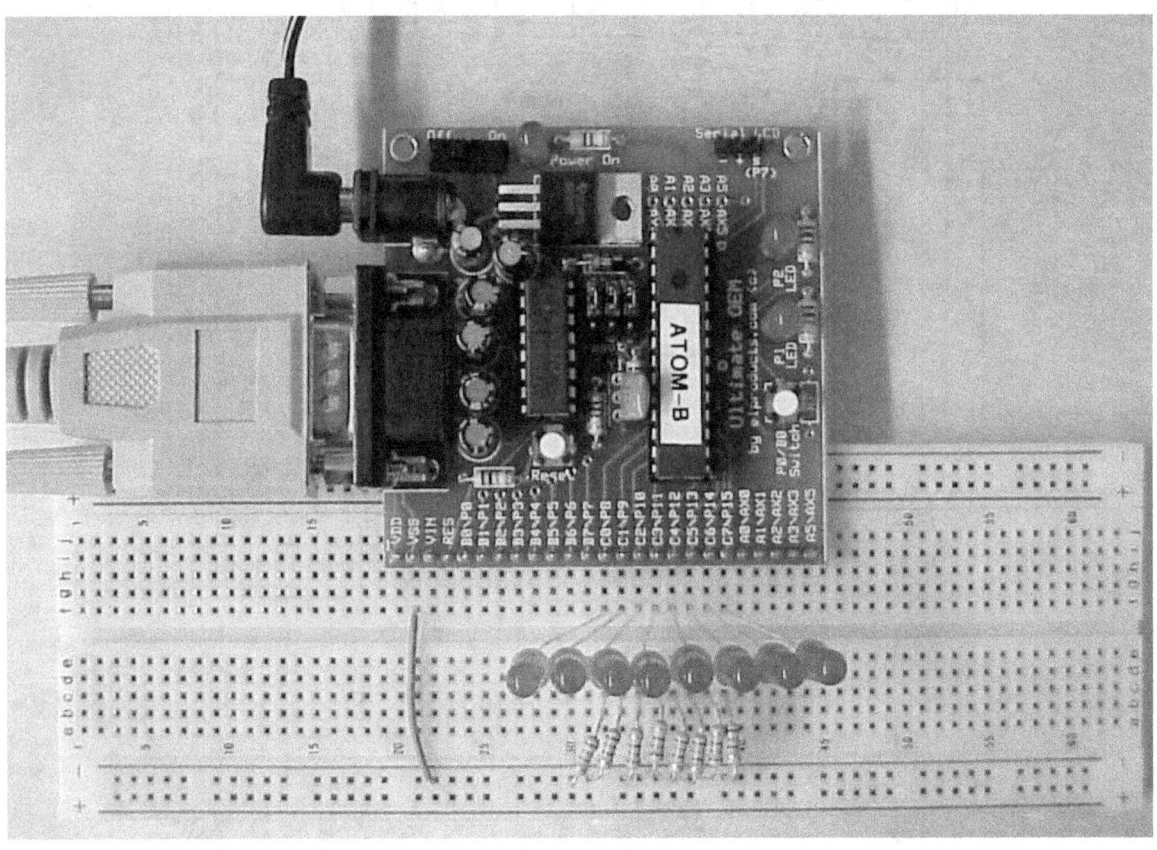

Project setup

The Ultimate OEM P8 thru P15 pins are connected to individual resistors and LEDs. The LEDs are grounded so a high signal at the I/O pin will light the LED and a low signal will turn the LED off.

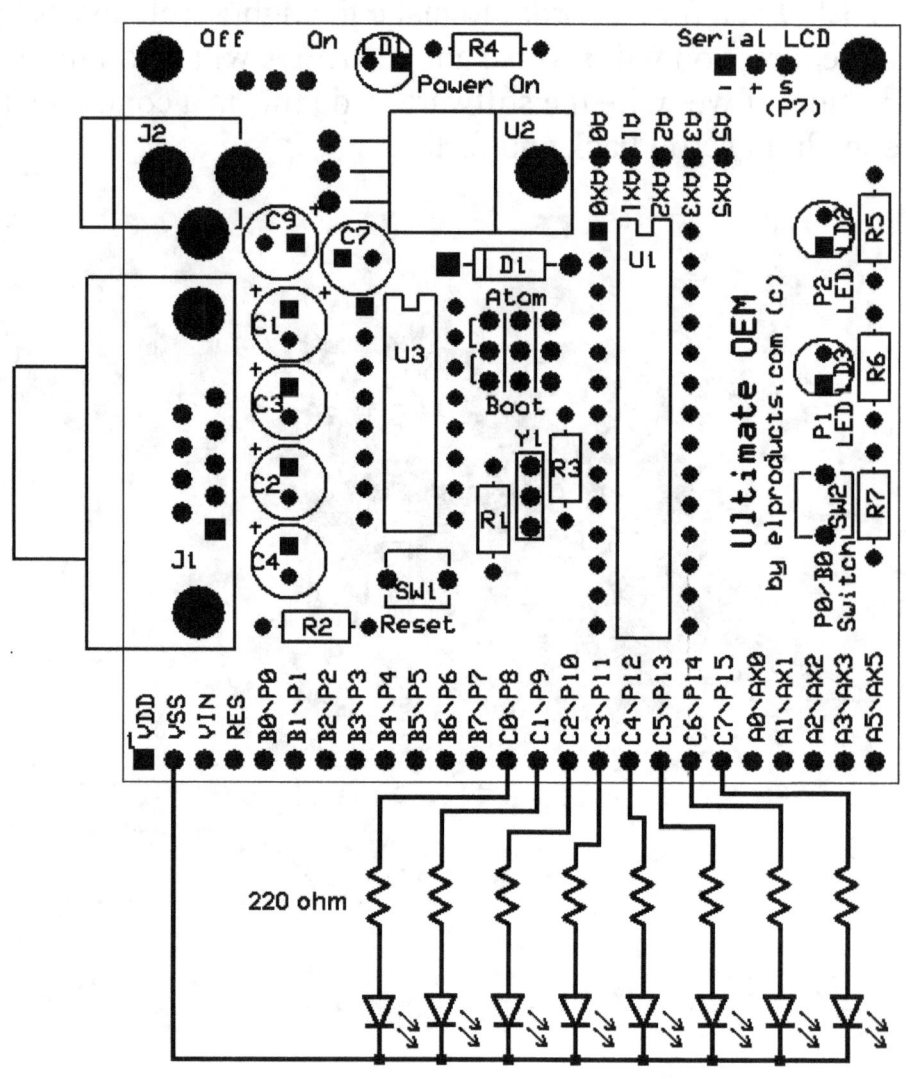

Software

The program listing below will perform the scroll LEDs function. You can type the program in or you can also download it off the included disk.

```
' *** Program Variables ***
x var byte              ' FOR-NEXT loop variable defined


' *** Main program loop ***

Main:                   ' Main loop label
for x = 8 to 15         ' Loop 8 times thru code
high x                  ' Turn on next LED
pause 10                ' Delay 10 milliseconds
low x                   ' Turn off LED
pause 10                ' Delay 10 milliseconds
next                    ' Is x = 7 yet?

for x = 15 to 8 step –1          ' Loop 8 times in negative direction
high x                  ' Turn next LED on
pause 10                ' Pause 10 milliseconds
low x                   ' Turn off LED
pause 10                ' Pause 10 milliseconds
next                    ' Is x = 0 yet?

goto main               ' Jump to main and do it again

end                     ' Stop if the program gets here
```

How it works

The software program controls the LEDs and flashes them one at a time in a continuous scrolling motion of light. To understand the software, lets step through the major sections of code.

The top section states: " *** Define Program Variables *****". This is where the variable gets defined. A single variable is established and given the simple nickname of "x". The program will use that later as a general storage space for the FOR-NEXT loop command counter.

Program variables need to have their size defined. In this example "x" does not need to be larger than 255 so a byte size will do. A variable can be defined as a bit, nibble (4 bits, 0-16), byte (8 bits, 0-255) or word (16 bits, 0-65535). You'll see that more in later projects.

```
' *** Program Variables ***
x var byte
```

After the variable definition, the main program loop is entered. The description line " '*** Main program loop ***" begins this section and it's quickly followed by the "main:" label.

```
' *** Main program loop ***

main:
```

This label defines a location, within the program, where the main section of code begins. Under this label is where the LEDs are functioned. The HIGH and LOW commands are once again used to turn the LED on (HIGH) and off (LOW). Because this program has to repeat the same function for each LED, the program could have a bunch of HIGH and LOW commands. That would work but would also take up more program memory and require lots of typing.

To simplify the program, the FOR-NEXT command will be used. The FOR-NEXT command creates a small loop where everything between the line that starts with FOR and the line that starts with NEXT is rerun a specified number of times. For example, look at the code section below.

```
for x = 8 to 15
high x
pause 10
low x
pause 10
next
```

The section of code starts with the FOR command followed by a simple little math type statement, "x = 8 to 15". What this means is, every time this list of commands is functioned, increment the variable x by one starting with 8 and ending at 15. This is known as a FOR-NEXT loop.

The value of the variable "x" is actually tested at the NEXT command line. If "x" equals 15 the program leaves the FOR-NEXT loop and moves on to the commands following the NEXT command line. If "x" does not equal 15 the program control jumps back to the FOR line and "x" is increased by 1. This allows the programmer to write a chunk of repeating commands without having to write them over and over again.

In this loop of code, notice how the HIGH and LOW commands do not use the P nicknames for the LEDs. Each time the program runs through another of the FOR-NEXT loops, a different LED should be lit. By using the variable "x" after the HIGH and LOW commands, a different LED is turned on and then off. The value of "x" must match up to the pins connected to the LEDs. In this case, because the LEDs are connected to P8 thru P15, we can use 8 thru 15 as the FOR-NEXT counter values while also using them as the pin variable in the HIGH or LOW commands.

This project uses a very short delay time between the HIGH and LOW commands. The intent is to make the light scroll forward quickly. Notice I only said forward. That's because this FOR-NEXT loop only lights LEDs 8 to 15. It does not reverse the direction and light 15 thru 8. That requires a separate FOR-NEXT loop.

The next section of code is very similar and is another FOR-NEXT loop. In this section though, the FOR-NEXT loop counts down instead of counting up.

```
for x = 15 to 8 step -1
high x
pause 10
low x
pause 10
next
```

That change of counting direction is done with the "step –1" added to the FOR command line. The "step –1" directs the FOR-NEXT command to add a "-1" (negative 1) to the value of "x" each time through the loop. The FOR command line starts at 15 and goes to 8. It's the opposite of the first FOR-NEXT loop.

The purpose of counting the opposite direction is to make the LED light scroll back from the far right LED to far left LED thus completing the back and forth scrolling motion.

After the FOR-NEXT loops are complete. The last command is the "GOTO MAIN" command line. It loops the program back up to the "main" label so it can run over and over again.

To make sure the program does not get lost, the END command is added at the bottom. If this command is ever encountered, the program will stop running until the reset switch is pressed or power is removed and reconnected.

Next steps

A simple change is to modify the delay times to slow down or speed up the LED scroll. You could also change the FOR – NEXT loops to scroll only the first four rather than all eight. You do that by changing the value "15" in the FOR command lines to "11".

Chapter 7 – Reading Switches

Description

This project demonstrates how to read switches to control a function. The four switches are wired to the P4 thru P7 Atom pins. LEDs are connected to Atom pins P8, P10, P12 and P14. When a switch is pressed an LED will light as long as the switch is held down. Once the switch is released, the LED will not light. Each switch is tied to a different LED. P4 switch is connected to the P8 LED, P5 switch to P10 LED, P6 switch to P12 LED and P7 switch to P14 LED. These connections are made through the software program.

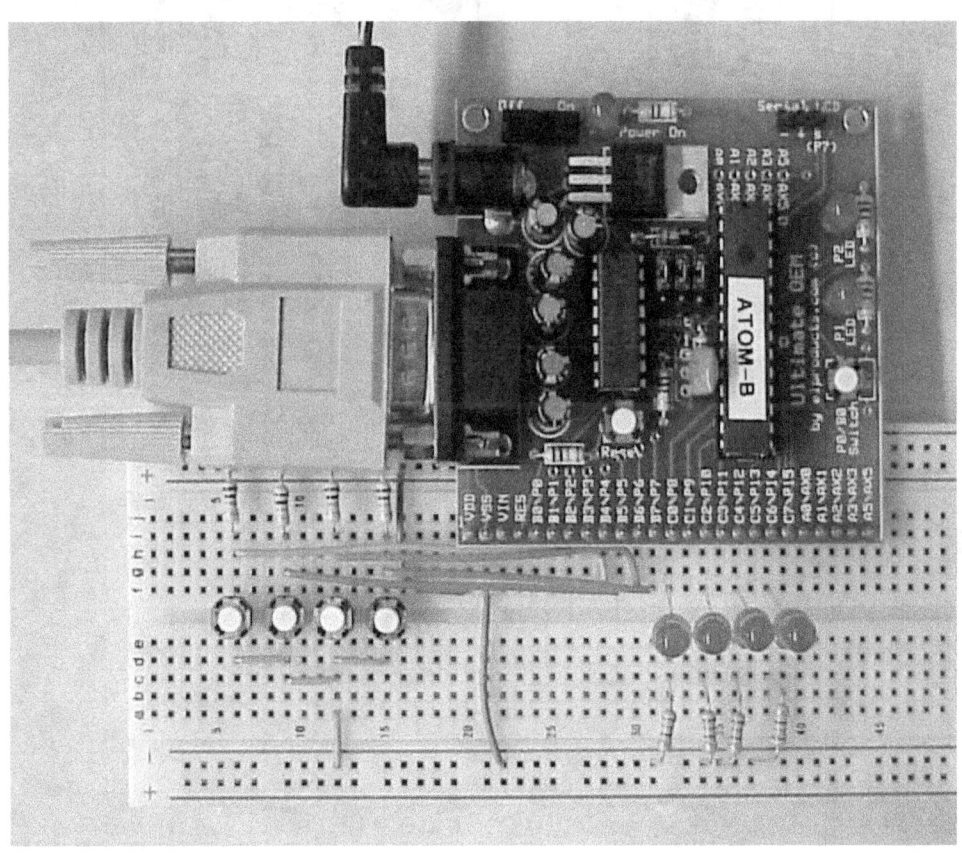

Project setup

The switches are wired as pull-down switches meaning they are normally open momentary switches connected to ground. The other end of the switch is connected to the Ultimate OEM Atom pin. To guarantee the software will see a high voltage when the switch isn't being pressed, a pull-up 10k resistor was added to each switch. The 10k resistor connects to a 5-volt common rail at one end and to the I/O pin at the other end. The 5-volt rail is supplied from the Vdd pin of the Ultimate OEM module. The schematic below shows the connections.

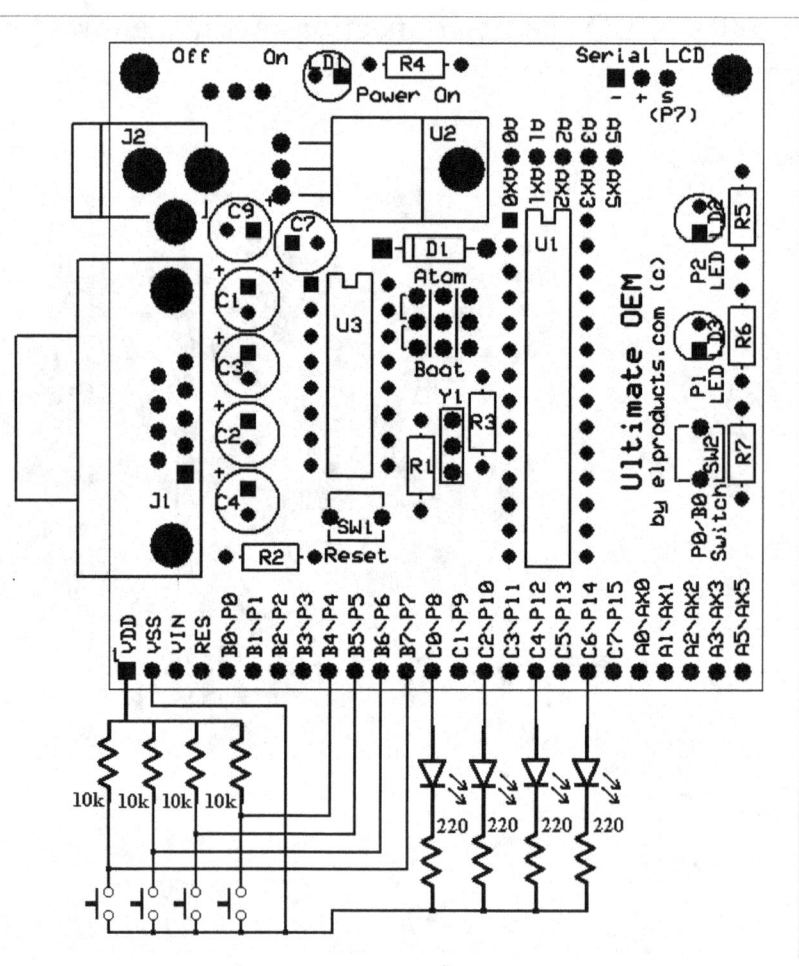

Software

The program listing below will perform the switch function. You can type the program in or you can also download it off the included disk.

```
"*** Define nicknames and Variables ***
sw1 var in4            ' Switch 1 connection nickname
sw2 var in5            ' Switch 2 connection nickname
sw3 var in6            ' Switch 3 connection nickname
sw4 var in7            ' Switch 4 connection nickname

led0 con 8             ' LED1 connection nickname
led1 con 10            ' LED2 connection nickname
led2 con 12            ' LED3 connection nickname
led3 con 14            ' LED4 connection nickname


"*** Main Program Loop ***
main
if sw1 = 0 then        'Check if switch 1 is pressed
high led1              'Light LED1 if pressed
elseif sw1 = 1         'Check if switch 1 is not pressed
low led1               'LED1 turned off
endif

if sw2 = 0 then        'Check if switch 2 is pressed
high led2              'Light LED2 if pressed
elseif sw2 = 1         'Check if switch 2 is not pressed
low led2               'LED2 turned off
endif

if sw3 = 0 then        'Check if switch 3 is pressed
high led3              'Light LED3 if pressed
elseif sw3 = 1         'Check if switch 3 is not pressed
```

```
low led3                'LED3 turned off
endif

if sw4 = 0 then         'Check if switch 4 is pressed
high led4               'Light LED4 if pressed
elseif sw4 = 1          'Check if switch 4 is not pressed
low led4                'LED4 turned off
endif

goto main               'Loop back to top and start over
```

How it works

The first part of the program defines the nicknames to make the program easier to read. In this case it also sets up the switch inputs for the software.

```
sw4 var in7             ' Switch 4 connection nickname
```

In order to use any pin as an input pin, the Atom software needs to be notified that you want to read it rather than control it. By putting the "in" in front of the Atom micro pin number and then using the "var" directive, the Atom software knows you want to treat the pin as an input and read the value present.

Rather than do this over and over again, the nickname above tells the Atom software, wherever the SW4 nickname is, treat it as reading the Atom micro pin 7 as an input.

Since we had a nickname for each switch, why not make a nickname for each LED? That is exactly what the line below does.

```
led1 con 8    ' LED1 connection nickname
```

Since we are driving the LEDs directly via Atom micro pins as outputs to either light the LED or turn it off, we don't need anything extra like the "in" token used above. The "con" directive is used for this since we are just renaming the P8 pin to LED1. Con is short for constant.

The main loop of the program contains four very similar routines. The first one is shown below. The first line tests the switch to see if it is pressed. The Atom micro pin sees a ground or zero voltage value ("0") if the switch is pressed.

if sw1 = 0 then

When a zero is measured, the next command is run. That command sets the Atom micro pin to a high or 5-volt level. This will light the first LED indicating the switch was pressed.

high led1

If the switch was not pressed, then the "high" command is skipped and the next command is issued. That command tests the switch again to see if it is at a high or 5 volt level ("1") rather than a zero.

elseif sw1 = 1

If the switch is not pressed the next command is run. That command turns the LED off by making the Atom micro pin low or at ground potential.

low led0

The last command just ends the series of IF-THEN tests and allows the software to continue on to the next set of commands.

endif

The rest of the program runs through the other blocks of code that test each of the four switches. If two switches are pressed at the same time, two LEDs should light up. The testing of the switches is part of a continuous loop and keeps checking the switches to see if anything has changed.

Someone reading this may ask; What if the switch is pressed in-between IF-THEN statements? Wouldn't this possibly be missed? The answer is; it's possible, but highly unlikely because the software in the Atom runs at about 30,000 instructions per second so a human would have to be really fast to press or release a switch in between IF-THEN statements.

Next steps
The LEDs that are controlled can be changed to light different LEDs rather than in order. This demonstrates how the software can change the function without having to rewire anything. You can also reverse the logic and have the LEDs light all the time unless a switch is pressed. That just requires reversing the High and Low commands or reversing the IF-THEN, ELSEIF values.

Because the switches are connected to Port B of the Atom micro, you can take advantage of the internal PIC pull-up resistors. These are only available on the P0 thru P7 pins. You enable it by issuing this command at the top of your program.

setpullup pu_on

This turns all the P0 thru P7 pullup resistors on. You cannot just turn a few on and a few off. It's all on or all off. To turn them off issue the "setpullup pu_off" command.

Chapter 8 – Driving a Relay

Description

This project is very simple but should be popular. Many people have asked how to control a relay with an Atom module. This project just controls an LED thru the relay contacts but it demonstrates how easy it is to control a relay with a an Ultimate OEM module or any Atom module.

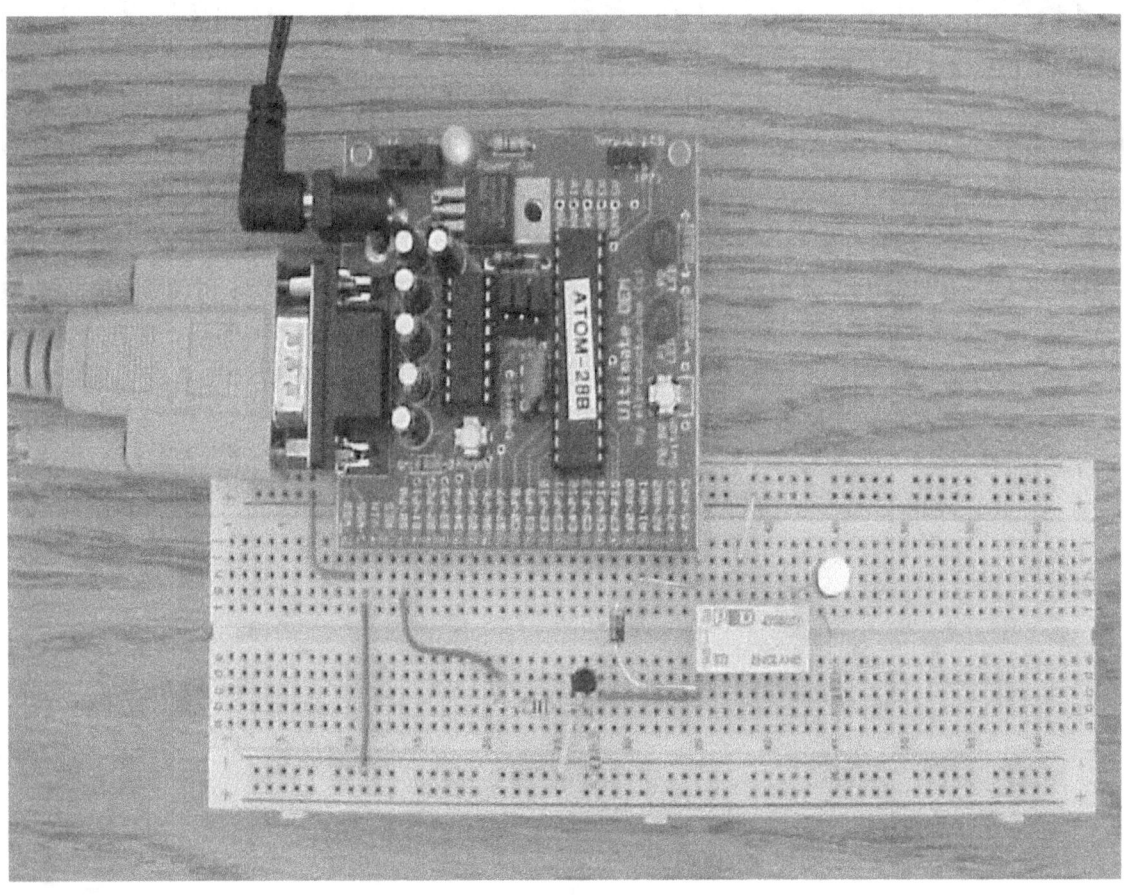

Project setup

The relay is wired to the Ultimate module through an NPN transistor. This is because the output pins of the Atom micro can only source or sink 25 milliamps maximum. Since the relay has a 125 ohm coil, it takes 40 milliamps to make it work. The 2N4401 can easily handle up to 300 ma so putting the transistor between the Ultimate I/O pin and the relay makes this very easy to control.

The diode across the coil of the relay helps protect the circuitry from inductive spikes. When the relay coil is turned off suddenly, the energy built up inside of it has to go somewhere. Without the diode, the energy would try to go into the transistor or create a magnetic interference in the Atom micro. The diode helps bleed the energy off safely.

The LED is powered from the same Vdd pin as the rest of the circuitry but in most cases you will power the relay control circuit from an outside supply. Many times a 12v or higher supply will be used. Because the relay is isolated from the Ultimate module thru the transistor, this shows a great way to control a high voltage item from a low voltage module. Always be careful though not to exceed the power capabilities of the relay contacts.

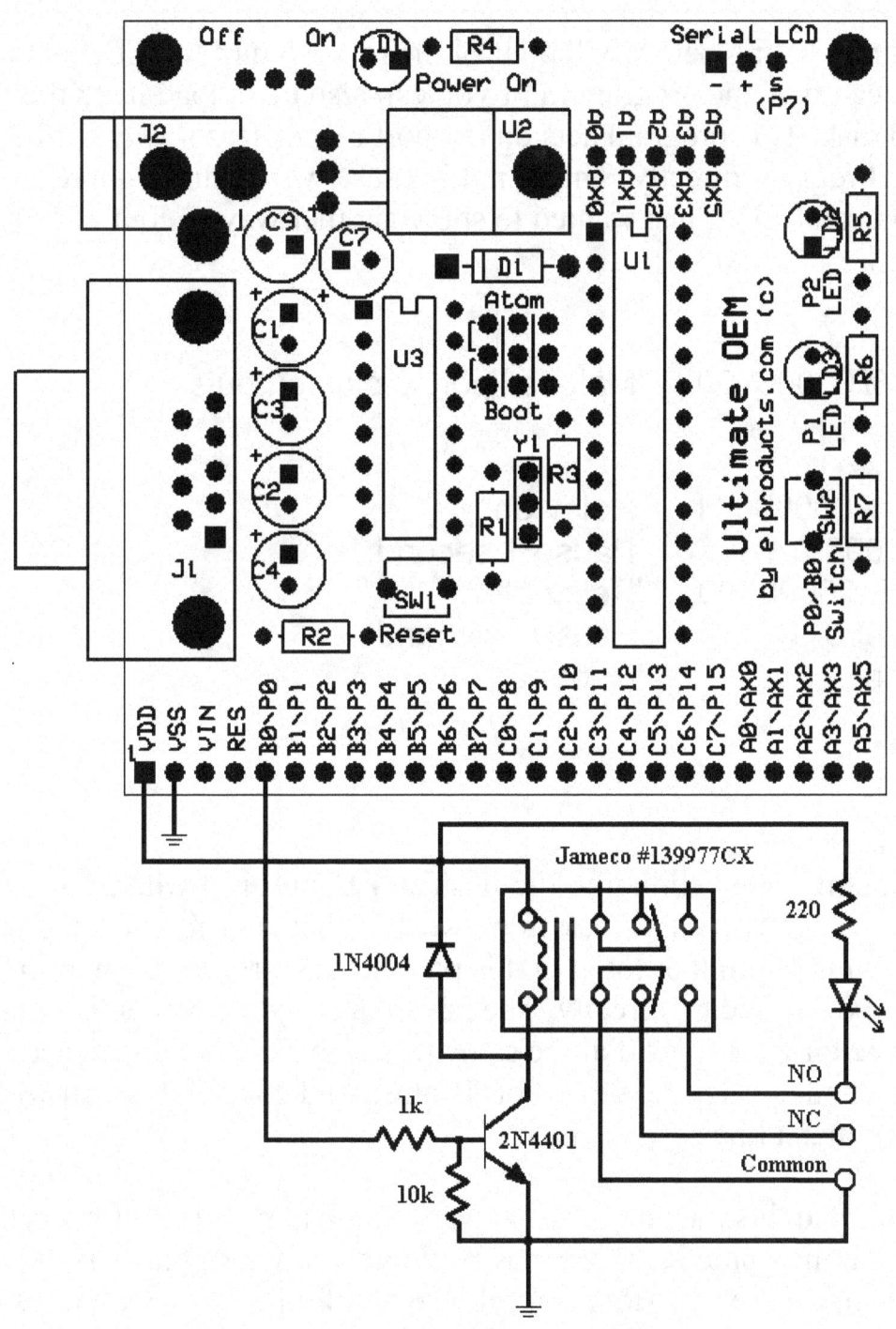

Software

The program listing below will control the relay to turn the LED on and off. You can type the program in or you can also download it off the included disk. This program acts on the port in a different way by driving the Port B register directly. This is not because we are driving a relay rather than an LED. I just wanted to show another way to control I/O with the Atom.

```
TrisB = %00000000     'Make all port B pins outputs

loop:
Portb = %00000001     'Relay on
pause 1000            'Pause 1 second
portb = %00000000     'Relay off
pause 1000            'Pause 1 second
goto loop             'Repeat
```

How it works

This program is very similar to the flash an LED project where we turn a port pin on and off with a delay in-between. What is different is the way the port pin is set high or low. In this project the port register within the Atom micro is acted on directly. The advantage to this method is multiple pins can be turned on or off at the same time. This method also makes it easy to add other relays to other Port B pins. And control them all with a single command line.

You won't find this method in the Atom command manual but it works and is a common practice in various Basic Compilers so this was a great way to demonstrate the Atom compiler some of the same capabilities other higher priced compilers do.

Pins P0 thru P7 are part of the PortB register of the Atom micro. Pins P8 thru P15 are part of the Port C register, so all this applies to PortC as well if you wanted to add a lot of relays.

The first part of the program does something very important, it establishes the Port B pins as outputs.

```
TrisB = %00000000     'Make all port B pins outputs
```

The name "TrisB" is a register in the Atom micro. The Atom compiler accepts that name and knows what you are modifying when this name is used. The Tris registers set the data direction as I explained earlier in the book. Each pin on Port B has a bit associated with it in the TrisB register. The left most bit is the P7 control, the right most is the P0 pin bit. A "1" makes the pin an input and a "0" makes the pin an output. In this example we set every bit to "0" so we are setting every pin in Port B to output mode.

Next we enter the main loop. We start with a simple label "loop" at the top and then we start to work on the Port B register directly.

```
loop:
Portb = %00000001     'Relay on
```

We use the "%" binary mode directive in front of the number so the Atom compiler knows we are giving it a binary number. A "0" makes the pin low (ground) and a "1" makes the pin high (5v). In the first line we set the P0 bit to a "1" which makes the port output a high level. This turns the transistor on and thus energizes the relay coil. This closes the relay contacts and turns on the LED connected to the relay. Then we issue a

PAUSE command to leave the relay closed and LED on for 1000 milliseconds or one second.

```
pause 1000              'Pause 1 second
```

In the next section we reverse it all, setting the P0 bit to "0" thus shutting off the transistor and in turn the relay. The LED shuts off and we delay this mode for one second.

```
portb = %00000000    'Relay off
pause 1000           'Pause 1 second
```

Finally we just jump back to the label "loop" and do it all again.

```
goto loop               'Repeat
```

You'll notice that I sometimes capitalize the name portb and sometimes I don't. I did this on purpose to show that the Atom compiler treats the capital letters and small letters the same.

When the project is running, you will not only see the LED turn on and off but you will hear the relay click. This can be added to a switch project if you need or want a clicking sound when the switch is pressed. It can also act as a turn signal sound like an automobile. This is very similar to the circuitry used in a car to turn the blinker lamp on and off.

Next steps
In the Atom command manual are the DIR, OUT and IN pin control names. DIR is similar to TRIS and OUT, IN are similar to what I did driving PortB directly. Try rewriting this project using those names to understand how they work the same. I personally like controlling the TRIS and Port B registers directly since this is the way a C or Assembly

language programmer does it. The DIR, OUT and IN are carryover from the Basic Stamp.

Another thing you can do is repeat the relay circuit but, connect it to a different Port B pin. Then you can control each relay separately. This can be a very handy project since a relay can control a lot of electrical items.

Chapter 9 – Liquid Crystal Display (LCD)

Description

This project is very simple but is a great starting point for all your future LCD projects. Having the opportunity to spell out information on an LCD module is a great addition to any project. This project setups a 2x16 LCD and then displays the advertisement "Ultimate OEM by elproducts.com". Then it waits a few 1/10ths of a second and displays it again. This is too many characters for one 16 character line so it is split between the first and second line as seen in the project PICtures.

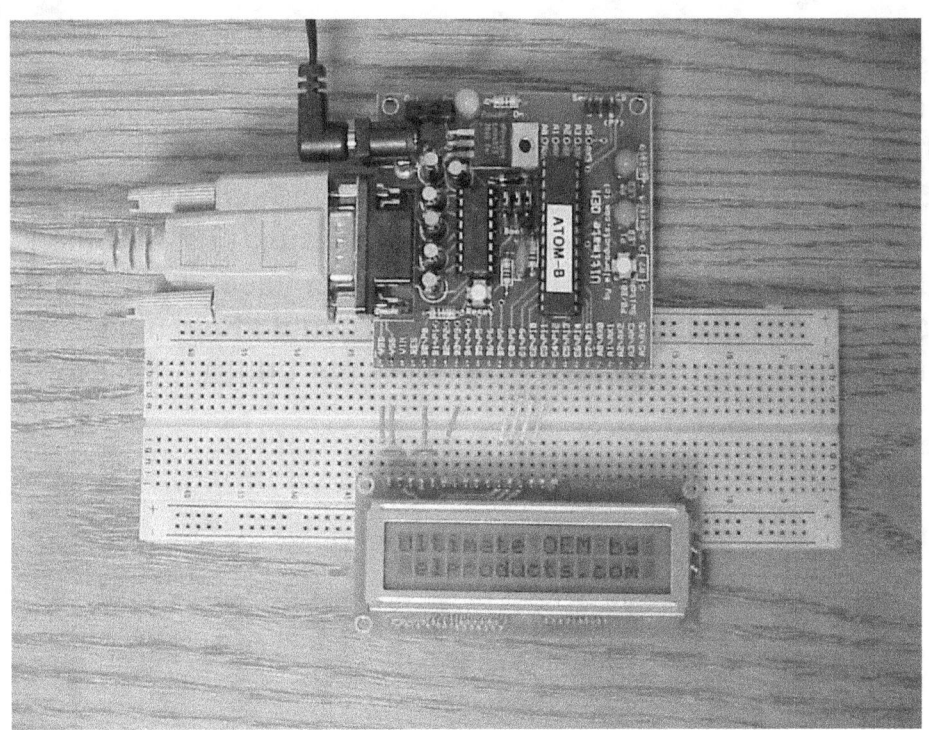

Close-up view of the LCD display

Project setup

The LCD plugs into the breadboard and then connections are made to the Ultimate OEM Atom module with jumpers per the schematic below. This project drives the LCD in 4 bit mode so DB0 through DB4 of the LCD remain unconnected. I also take a shortcut and connect the dimming input (Vo) to ground for maximum brightness. This could be replaced with a potentiometer circuit to adjust the brightness but I'll leave that for you to add if you wish.

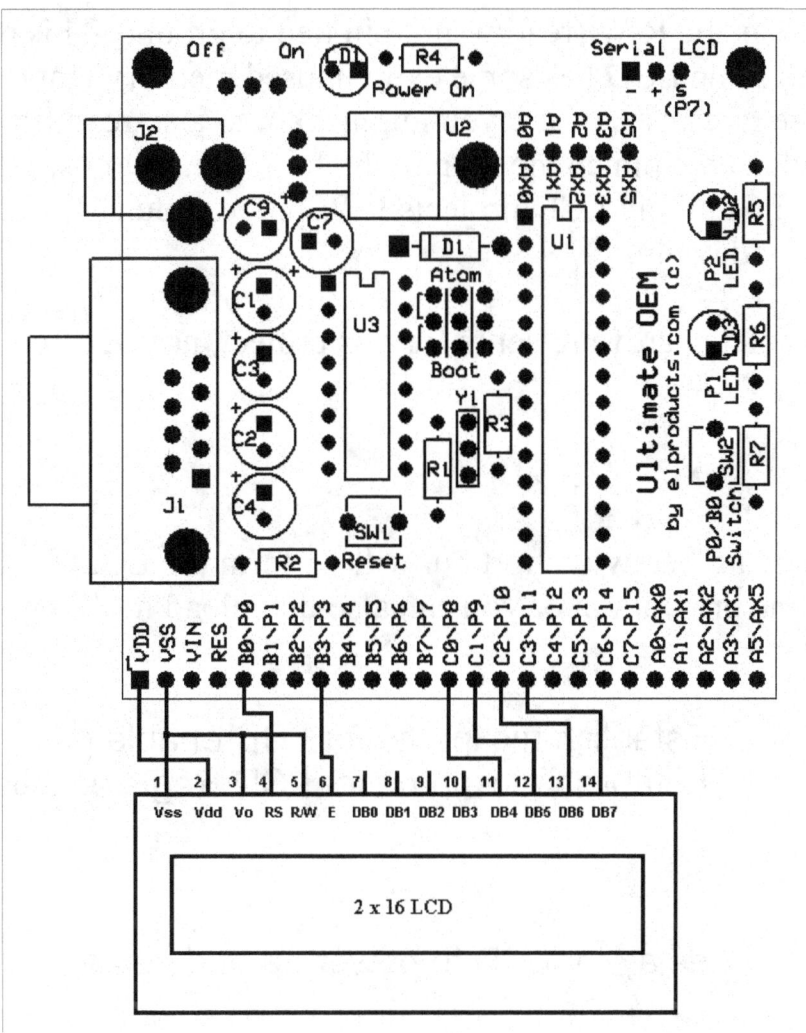

The LCD requires 6 connections from the Atom micro. The RS line, E line and the DB4-DB7 data lines make up those connections.

The Atom software makes driving an LCD very easy with the LCDWRITE command. In order to use that command though you need to know what port pins the LCD is connected to. In the schematic the RS line is on P0, E on P3, DB4 on P8, DB5 on P9, DB6 on P10 and DB7 on P11.

The LCD also has the R/W (read/write) pin tied to ground. This makes the LCD write only. The LCD has some extra unused memory that can be read so in those rare cases where you might want to also read from the LCD you would add a pin connection to the R/W pin. I almost never use that feature so I tie all my LCD projects R/W pin to ground to put the LCD in write mode only.

This is the complete hardware setup for this project, now lets move to the software.

Software

The program listing below is short but will perform the LCD function. You can type the program in or you can also download it off the included disk

```
epin con 3          'Establish nickname for LCD enable pin
rspin con 0         'Establish nickname for LCD Register Select pin



pause 500           'delay for LCD to power up and settle

' *** Initialize and setup 2x16 LCD as two line display *******

lcdwrite rspin\epin,outc,[initlcd1,initlcd2,twoline,scrblk,clear,home]

main

' ****** Display message on line 1 and line2 of LCD ******

lcdwrite rspin\epin,outc,[clear,home,"Ultimate OEM by",|
scrram+$40, " elproducts.com "]
```

```
pause 100  'delay for 100 milliseconds to steady display

goto main   'Repeat it again
```

How it works

As usual, the software for this project is very short and simple but forms a great building block for future programs you may want to write. It's actually the simplicity of the Atom software that makes these programs so short.

The heart of this program is the LCDWRITE command. The command requires two steps to make the LCD work. First the LCD has to be configured and the LCDWRITE command will do that. Second the information to be displayed has to be written out. The LCDWRITE command does this also.

At the top of the program two nicknames are established using the constant or "CON" directive. This makes it easier to setup the LCDWRITE command.

```
epin con 3        'Establish nickname for LCD enable pin
rspin con 0       'Establish nickname for LCD Register Select pin
```

The program follows with a "PAUSE 500" command line. All LCDs require time to power up before accepting any signals from the micro. Because the Ultimate OEM boots up so much faster than the LCD this delay of 500 milliseconds is enough time to let the LCD catch up.

The next line is the configuration line and it's shown below.

```
lcdwrite rspin\epin,outc,[initlcd1,initlcd2,twoline,scrblk,clear,home]
```

The LCDWRITE command first requires the RS and E pins to be defined. The nicknames we created get used here. We could have just used the digits 0 and 3 but it's usually easier to understand the program when descriptive nicknames are used.

Following those two, the data pins are defined. The data pins are grouped together as 4 consecutive I/O pins. This is no accident that I connected those in series. The LCDWRITE command requires that. The "OUTC" entry is a nickname that represents P8 through P11 that are connected to the LCD.

The last part of the command is in brackets. This portion has control codes the LCD uses for configuration. Most of these are required but some are optional. The "initlcd1" and "initlcd2" are required and send the necessary command codes to the LCD.

The next line "twoline" is a code to configure the LCD to use both the top and bottom lines of the display. The "scrblk" sets the display to on and the blinking block to on but turns off an underline cursor. Finally the optional "clear" and "home" are included. These position the cursor to the first LCD position and clear anything on the display. All the options for the LCDWRITE command are in the Atom software manual.

The "main" program loop uses the same LCDWRITE command to display "Ultimate OEM by elproducts.com". The same rspin\epin and outc are used and the "clear" and "home" are included inside the brackets to position the cursor at the beginning of the first line. That should all be familiar.

```
lcdwrite rspin\epin,outc,[clear,home,"Ultimate OEM by", scrram+$40, " elproducts.com "]
```

The characters that are listed between the quote marks are the characters to be displayed on the LCD. This line was long so I had to shrink the font to fit on book a page. After the "clear,home," is the "Ultimate OEM by" characters. These will be displayed at the first character block and continue to the right from there. The characters in the quotes have to be equal to or less than 16 characters to fit. If you put more than 16 on a 2x16 display, the last characters will not be displayed. In other words, they will not roll over to the next line because that is not how a 2x16 display works. To place characters on the second line, you have to move the cursor to the beginning of the second line and that is at hex value $40.

You see the 44780 chip that drives the LCD is a 2x40 display chip. It drives all LCDs like they are a 2 line 40 character display. Most LCD modules though are built based on a 4x20 arrangement where the third line is just a continuation of the first line as seen in the block layout below. Each value in the blocks is the hexadecimal value for that position.

00	01	02	03	04	05	06	07	08	09	10	11	12	13	14	15	16	17	18	19	← Character position (dec.)
00	01	02	03	04	05	06	07	08	09	0A	0B	0C	0D	0E	0F	10	11	12	13	← Row0 DDRAM address (hex)
40	41	42	43	44	45	46	47	48	49	4A	4B	4C	4D	4E	4F	50	51	52	53	← Row1 DDRAM address (hex)
14	15	16	17	18	19	1A	1B	1C	1D	1E	1F	20	21	22	23	24	25	26	27	← Row2 DDRAM address (hex)
54	55	56	57	58	59	5A	5B	5C	5D	5E	5F	60	61	62	63	64	65	66	67	← Row3 DDRAM address (hex)

In this project we are using a 2x16 LCD so the top two lines of the 4x20 arrangement form the two lines of the 2x16 display. The two lines start at "00 hex" and "40 hex" but they end earlier than shown. The first line ends at "0F hex" and the second ends at "4F hex". In other words the first line actually stops at the 15th column. If you were to send 18 characters to the first line instead of 16, the last two would not show up because they are off the screen at the 16th and 17th column that are not included in a 2x16 display. This is why the scrram+$40 precedes the "elproducts.com" in the LCDWRITE command so it shows up in the second line.

The SCRRAM is an LCDWRITE command setup code to position the cursor at a specific display memory location. Since we want to start at the beginning of the 2nd line, we have to refer to the layout above and place it at 40 hex or $40 in Atom speak. When we use the directive HOME in the LCDWRITE command, it is the same as SCRRAM + $0.

After the LCDWRITE command, the program just delays 100 milliseconds using the PAUSE command and loops back to send "Ultimate OEM by elproducts.com" again. We really don't need to loop because the LCD will continue to display what is in its memory until power is removed but looping is just a normal way to write code.

That is all it takes to control an LCD display with the Atom.

Next steps

The most obvious next step is to change the characters between the quotes to display something other than my crummy commercial display of "Ultimate OEM by elproducts.com ". Further next steps require the user to read more about the LCDWRITE command and all the command codes available. I list them with the LCDWRITE command in Chapter 3. You can also drive a 4x20 display like we do in the next chapter.

Chapter 10 – Large Digits on 4x20 LCD

Description

This project continues where the previous project left off. We drive a 4x20 LCD with the Ultimate OEM Atom module but also send custom characters to the LCD's character memory so we can make custom large characters and numbers. The project here will count using the hexadecimal number system so it will count and display the numbers 0 – 9 and A - F in large character format.

The custom characters we use to build the large characters are shown in the close-up PICture of the display below. These custom characters are stored in the LCD memory so we can call them up the same way we would call up any other ASCII character.

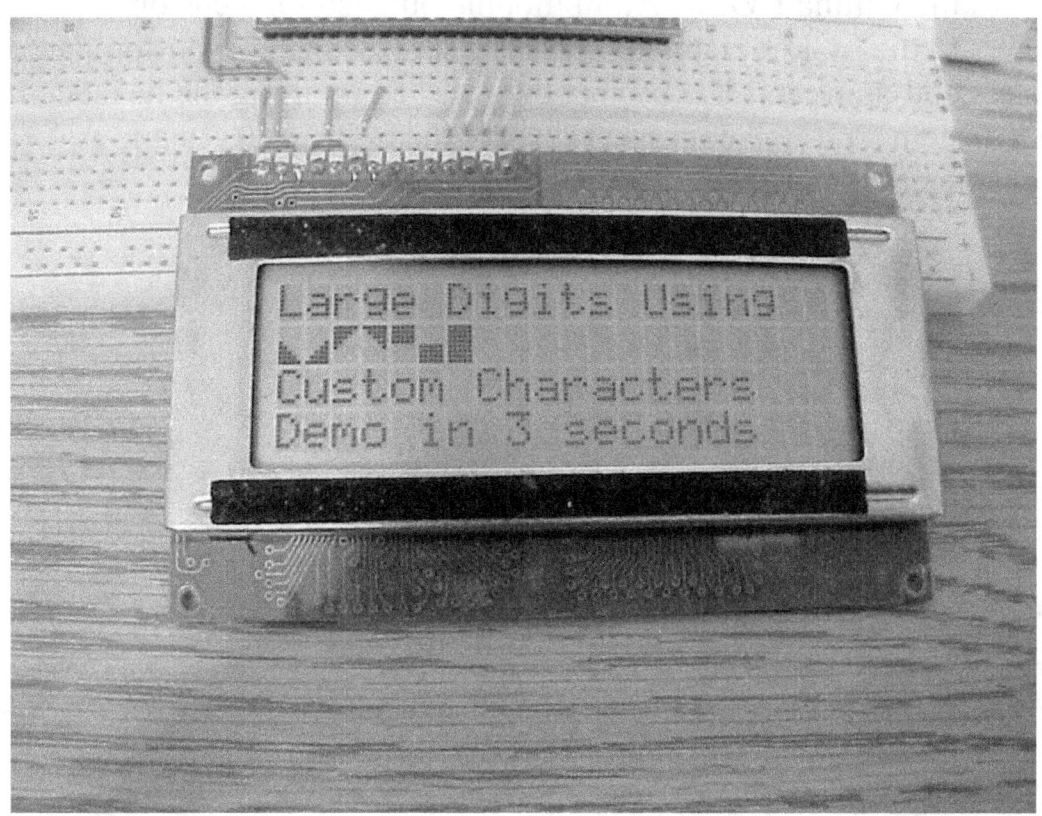

The number "1" is displayed below using the custom characters. It makes it very easy to read numbers of this size from across the room.

Project setup

The LCD plugs into the same connections as the previous project. A great advantage to using LCD modules is their common connection system which makes it easy to change from 2x16 to 4x20. The connections for the 2x16 used in the previous project are just carried over to this project.

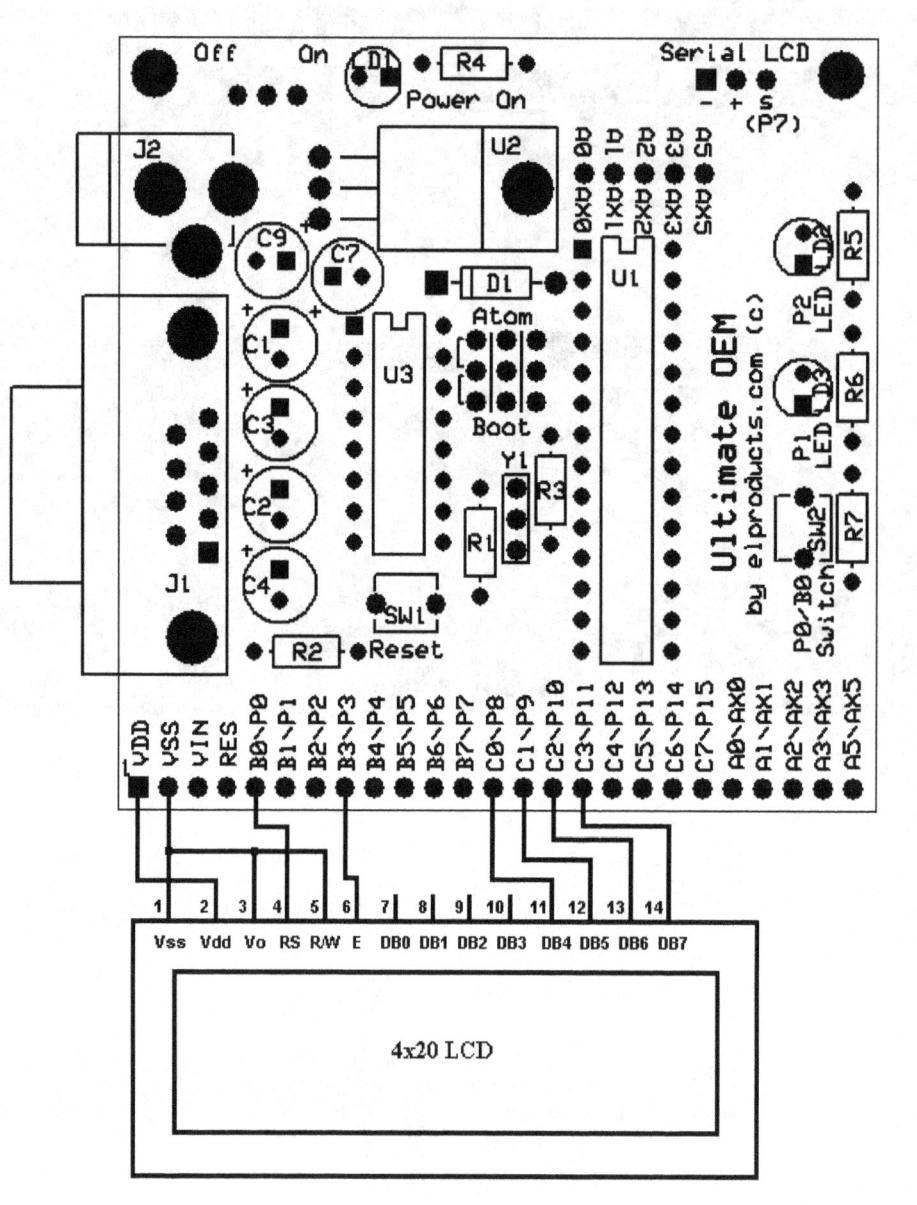

Software

This software is much bigger than the 2x16 example but most of the code deals with setting up the LCD memory to display custom characters. In the code you will notice the "|" pipe character at the end of several lines. This is for line continuation. This is a special character that the Atom compiler recognizes as a continuation message. When the compiler sees that character, it knows the command line was too long for the editor window and continues on the next line. Setting up the characters takes a lot of space so the line continuation function is used often. This is a special code you won't find in the Atom manual. At least I couldn't find it. I learned of it directly from Basic Micro.

```
x var byte
char var byte
epin con 3            'Establish nickname for LCD enable pin
rspin con 0           'Establish nickname for LCD Register Select
pin

' *** Initialize LCD ***
pause 500
lcdwrite rspin\epin,outc,[initlcd1,initlcd2,twoline,scr,clear,home]

'*** Create Custom Characters in LCD memory locations 0-7 ***

lcdwrite rspin\epin,outc,[CGRAM]
for x = 0 to 63

lookup x,[$00,$00,$00,$10,$18,$1C,$1E,$1F,$00,$00,$00,$01,|
$03,$07,$0F,$1F,$1F,$1E,$1C,$18,$10,$00,$00,$00,$1F,$0F,|
$07,$03,$01,$00,$00,$00,$1F,$1F,$1F,$1F,$00,$00,$00,$00,|
$00,$00,$00,$00,$1F,$1F,$1F,$1F,$1F,$1F,$1F,$1F,$1F,$1F,|
$1F,$1F,$00,$00,$00,$00,$00,$00,$00,$00], char
```

```
lcdwrite rspin\epin,outc,[char]
next

' *** Initial screen with program description ***
main
lcdwrite rspin\epin,outc,[clear,home,scrram,"Large Digits Using"]
lcdwrite rspin\epin,outc,[scrram + $40]

for x = 0 to 7
lcdwrite rspin\epin,outc,[x]
next
lcdwrite rspin\epin, outc,[scrram + $14, "Custom Characters"]
lcdwrite rspin\epin, outc,[scrram + $54, "Demo in 3 seconds"]
pause 3000

' *** LCD control code to display 0 - F large characters ****

' *** "0" character
lcdwrite rspin\epin,outc,[clear,home,scrram+$09,6,4,6,scrram+$49,|
6,7,6,scrram+$1D,6,7,6,scrram+$5D,6,5,6]
pause 1000

' *** "1" character
lcdwrite rspin\epin,outc,[clear,home,scrram+$09,1,6,scrram+$4A,|
6,scrram+$1E,6,scrram+$5D,5,6,5]
pause 1000

' *** "2" character
lcdwrite rspin\epin,outc,[clear,home,scrram+$09,4,4,6,scrram+$49,|
5,5,6,scrram+$1D,6,7,7,scrram+$5D,6,5,5]
pause 1000
```

```
' *** "3" character
lcdwrite rspin\epin,outc,[clear,home,scrram+$09,4,4,6,scrram+$49,|
5,5,6,scrram+$1D,7,7,6,scrram+$5D,5,5,6]
pause 1000


' *** "4" character
lcdwrite rspin\epin,outc,[clear,home,scrram+$09,6,7,6,scrram+$49,|
6,7,6,scrram+$1D,4,4,6,scrram+$5D,7,7,6]
pause 1000


' *** "5" character
lcdwrite rspin\epin,outc,[clear,home,scrram+$09,6,4,4,scrram+$49,|
6,5,5,scrram+$1D,7,7,6,scrram+$5D,5,5,6]
pause 1000


' *** "6" character
lcdwrite rspin\epin,outc,[clear,home,scrram+$09,6,4,4,scrram+$49,|
6,5,5,scrram+$1D,6,7,6,scrram+$5D,6,5,6]
pause 1000


' *** "7" character
lcdwrite rspin\epin,outc,[clear,home,scrram+$09,4,4,6,scrram+$49,|
7,7,6,scrram+$1D,7,7,6,scrram+$5D,7,7,6]
pause 1000


' *** "8" character
lcdwrite rspin\epin,outc,[clear,home,scrram+$09,6,4,6,scrram+$49,|
6,5,6,scrram+$1D,6,4,6,scrram+$5D,6,5,6]
pause 1000


' *** "9" character
lcdwrite rspin\epin,outc,[clear,home,scrram+$09,6,4,6,scrram+$49,|
6,5,6,scrram+$1D,7,7,6,scrram+$5D,7,7,6]
```

```
pause 1000

' *** "A" character
lcdwrite rspin\epin,outc,[clear,home,scrram+$09,6,4,6,scrram+$49,|
6,5,6,scrram+$1D,6,7,6,scrram+$5D,6,7,6]
pause 1000

' *** "b" character
lcdwrite rspin\epin,outc,[clear,home,scrram+$09,6,7,7,scrram+$49,|
6,5,5,scrram+$1D,6,7,6,scrram+$5D,6,5,6]
pause 1000

' *** "C" character
lcdwrite rspin\epin,outc,[clear,home,scrram+$09,6,4,4,scrram+$49,|
6,7,7,scrram+$1D,6,7,7,scrram+$5D,6,5,5]
pause 1000

' *** "d" character
lcdwrite rspin\epin,outc,[clear,home,scrram+$09,7,7,6,scrram+$49,|
5,5,6,scrram+$1D,6,7,6,scrram+$5D,6,5,6]
pause 1000

' *** "E" character
lcdwrite rspin\epin,outc,[clear,home,scrram+$09,6,4,4,scrram+$49,|
6,5,5,scrram+$1D,6,7,7,scrram+$5D,6,5,5]
pause 1000

' *** "F" character
lcdwrite rspin\epin,outc,[clear,home,scrram+$09,6,4,4,scrram+$49,|
6,5,5,scrram+$1D,6,7,7,scrram+$5D,6,7,7]
pause 1000

' **Final message from program before looping back to the top **
```

```
lcdwrite rspin\epin,outc,[clear,home,scrram,"Just imagine what",|
scrram+$40,"you can do!"]
pause 3000
goto main
```

How it works

Before we go through this code lets explain how the custom character
setup in the LCD works. The LCD has eight locations at the beginning of
its character memory that can be modified to make custom characters.
Once those are setup the custom characters can be called the same way
any standard pre-stored ASCII character can.

In the PICture on the next page, the 5x7 character bit map is shown with
the first custom character the program will define.

Custom LCD Character using the "Character Box"

Character Box

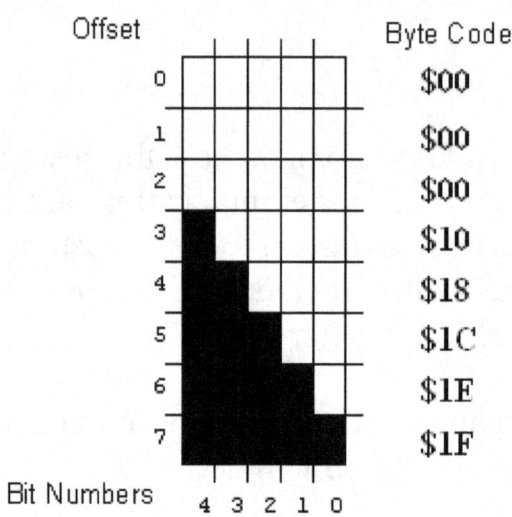

Offset

	Byte Code
0	$00
1	$00
2	$00
3	$10
4	$18
5	$1C
6	$1E
7	$1F

Bit Numbers   4 3 2 1 0

This is the custom character we place at memory location $00 of the LCD character memory. It forms a small ramp that slopes downward.

Each row of the character is defined by a byte value. Since the characters are only 5 bits wide in size, the three most significant bits of the byte value are always zero. The highest bit value is the 5th bit (bit number 4). When it is set, a block appears at that location. If you look at the fourth line of the character box (offset row value 3), the byte to the right shows $10 or binary %00010000. This makes the 5th block solid black but the rest of the row clear. The next line sets two blocks black by using byte $18 or binary %00011000. By setting the proper bits we can create any custom 5x8 character, which is what this program does. Let's go through the code.

First we establish a few variables and constants. The variables are just temporary storage locations labeled X and Char. The constants define the LCD "E" pin and "RS" pin.

```
x var byte
char var byte
epin con 3          'Establish nickname for LCD enable pin
rspin con 0         'Establish nickname for LCD Register Select pin
```

The program initializes the LCD by first waiting ½ second for it to warm up and then issues the LCDWRITE command to set it up as a 2 line LCD. If you remember in the last LCD project that the 3rd line of the LCD just continues at the end of the first line. Therefore a 4x20 display is really just a 2x40 display laid out differently.

Basic Atom actually has an LCDINIT command and that can be used in place of this but originally my method was the only way to initialize the LCD. It still works so I use it here to show what the manual doesn't show you. This way you only have to remember one LCD command not two.

```
' *** Initialize LCD ***
pause 500
lcdwrite rspin\epin,outc,[initlcd1,initlcd2,twoline,scr,clear,home]
```

The next section is the heart of this program. In this block of code the custom characters are created and stored in the LCD character memory locations 0 thru 7. Each character takes 8 bytes of data for total of 64 bytes (8 characters times 8 bytes).

To do this we first have to point to location zero of the Character RAM. We do this with the LCDWRITE command again by sending the "CGRAM" pointer. We don't have to add an address value since it defaults to the zero or first location.

'*** Create Custom Characters in LCD memory locations 0-7 ***

lcdwrite rspin\epin,outc,[CGRAM]

Now we send the custom characters to the RAM by using a FOR-NEXT loop and the LOOKUP command. The FOR-NEXT loop counts from 0 to 63 for a total of 64 loops and it defaults to stepping one count per loop. The variable "x" stores the present loop count value. The LOOKUP command then takes the value of "x" and jumps that many places, reads the byte value and stores it in the "char" variable. For example, lets assume x = 5 or the 6th time through the loop. The value of "char" will equal $1C since it is the sixth value listed.

for x = 0 to 63

lookup x,[$00,$00,$00,$10,$18,$1C,$1E,$1F,$00,$00,$00,$01,|
$03,$07,$0F,$1F,$1F,$1E,$1C,$18,$10,$00,$00,$00,$1F,$0F,|
$07,$03,$01,$00,$00,$00,$1F,$1F,$1F,$1F,$00,$00,$00,$00,|
$00,$00,$00,$00,$1F,$1F,$1F,$1F,$1F,$1F,$1F,$1F,$1F,$1F,|
$1F,$1F,$00,$00,$00,$00,$00,$00,$00,$00], char

lcdwrite rspin\epin,outc,[char]
next

The custom characters are now loaded. We can use them to create the large characters on the LCD. The "main" label starts the central program loop. In the section below "main" we use the LCDWRITE command to display a description of what this program will do. This was displayed in the 2nd PICture shown at the beginning of the chapter. We display "Large Digits Using" by using the LCDWRITE command.

```
' *** Initial screen with program description ***
main
lcdwrite rspin\epin,outc,[clear,home,scrram,"Large Digits Using"]
lcdwrite rspin\epin,outc,[scrram + $40]
```

In this section we call up the custom characters one at a time using a FOR-NEXT loop and the LCDWRITE command. The variable "x" holds a value from 0 to 7. LCDWRITE directs the LCD to display characters 0 thru 7, which are the characters we created above. See how easy it is to display custom characters once they are created?

```
for x = 0 to 7
lcdwrite rspin\epin,outc,[x]
next
```

We finish this block of code by displaying "Custom Characters" and "Demo in 3 seconds" to the display lines 3 and 4. SCRRAM +$14 is the beginning of line 3 and SCRRAM +$54 is the beginning of line 4.

```
lcdwrite rspin\epin, outc,[scrram + $14, "Custom Characters"]
lcdwrite rspin\epin, outc,[scrram + $54, "Demo in 3 seconds"]
```

Finally we pause 3 seconds so you can read the display and then move on to the next section.

```
pause 3000
```

From here the program creates the custom large characters using the custom small characters stored in CGRAM. I'll just describe the digit "1" shown in the PICture earlier since all the other large character sections below operate the same.

The #1 character is created by placing custom characters 1 and 6 on line one, character 6 on line two, character 6 on line three and characters 5,6 and 5 on line four. We pause 1 second so the digit can be read. The scram is offset with values that center the "1" on the LCD.

```
' *** "1" character
lcdwrite rspin\epin,outc,[clear,home,scrram+$09,1,6,scrram+$4A,|
6,scrram+$1E,6,scrram+$5D,5,6,5]
pause 1000
```

This is repeated again for all the hexadecimal characters from 0 thru F. Each is displayed and then a final message is displayed. The final section of code displays "Just imagine what you can do".

```
' *** Final message from program before looping back to the top ***

lcdwrite rspin\epin,outc,[clear,home,scrram,"Just imagine what",|
scrram+$40,"you can do!"]
pause 3000
goto main
```

With this custom character method, you can create just about anything on an LCD screen.

Next steps
The projects that can result from this are endless. The thing to remember is nothing stops you from redefining the custom characters in the middle of the program. For example, let's say you want to display large characters initially and then later in the program want to create an animation using different custom characters.

After completing the large custom number characters, clear the LCD screen and then load new custom characters in CGRAM locations 0-7. From these new characters, you can create the animation. Since the custom characters load in CGRAM so fast, the person watching the display just notices a frame change from words to large digits to animation. I've seen custom characters that had the old Pacman character eating dots across the screen.

Chapter 11 – Reading a Light Sensor

Description

This project introduces the Analog to Digital (A/D) converter port built into the Atom micro. The project reads a CDS light sensor to measure the level of ambient light in the room. A CDS light sensor changes its resistance based on the amount of light it is exposed to. This project uses the Atom micro's A/D port AX1 to convert the light sensor resistance value to a digital value that can be used in the Atom program. This project will display the level of light on an LCD as a digital value from 0 to 1024. A decision will also be made to turn on an LED at a set threshold so this becomes a light monitor.

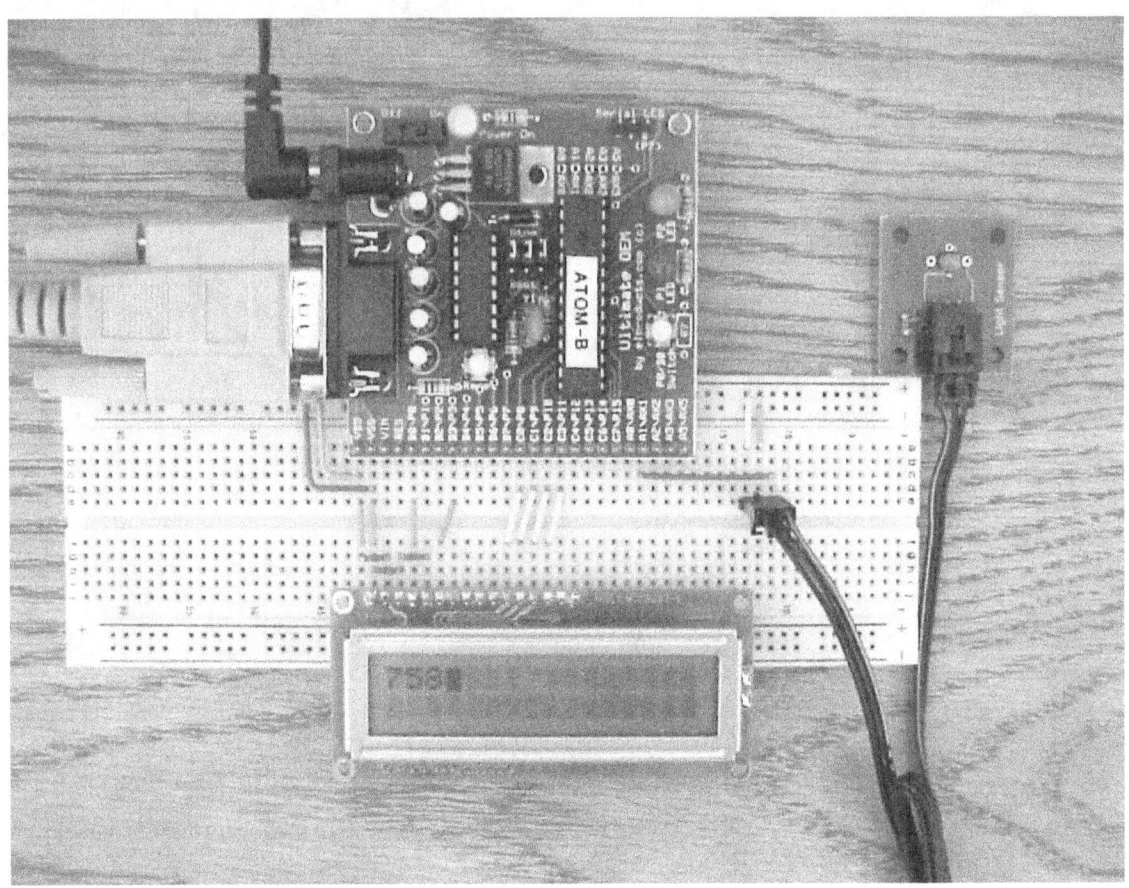

Light Sensor

Project setup

This project uses the same LCD setup as Chapter 9 so if you saved that setup on a breadboard, you can re-use it here. One of the reasons I like the Ultimate OEM module is the ability to easily plug into a low cost breadboard. I can have several completed projects built up on breadboards and just move the Ultimate OEM module around.

The light sensor for this project is built on a separate circuit board. As the project schematic shows, it's just a resistor divider with a 10k resistor pull-up and the CDS cell as the pull-down. The signal is tapped from the center connection. With 5v applied to the pull-up resistor and ground at the pull-down CDS cell, a voltage will be produced based on the ambient light.

The sensor board output voltage is what the AX1 A/D port reads and converts to digital. This can easily be built directly on the breadboard with a CDS cell and potentiometer as the pull-up resistor so you can adjust the range. In fact that may be a better way to do this because of the adjustment capability. Based on my first book though, many readers wanted sources for modular components since many people just want to program and not fool with all the hardware. Therefore when I found this sensor I thought this was a great way to test it out and give the reader an easy way to get a "ready to use" sensor.

The LED used is built into the Ultimate OEM module. It is connected to the P1 pin thru a 220 ohm resistor. This can also be done on the breadboard similar to the LED in the Chapter 5 project.

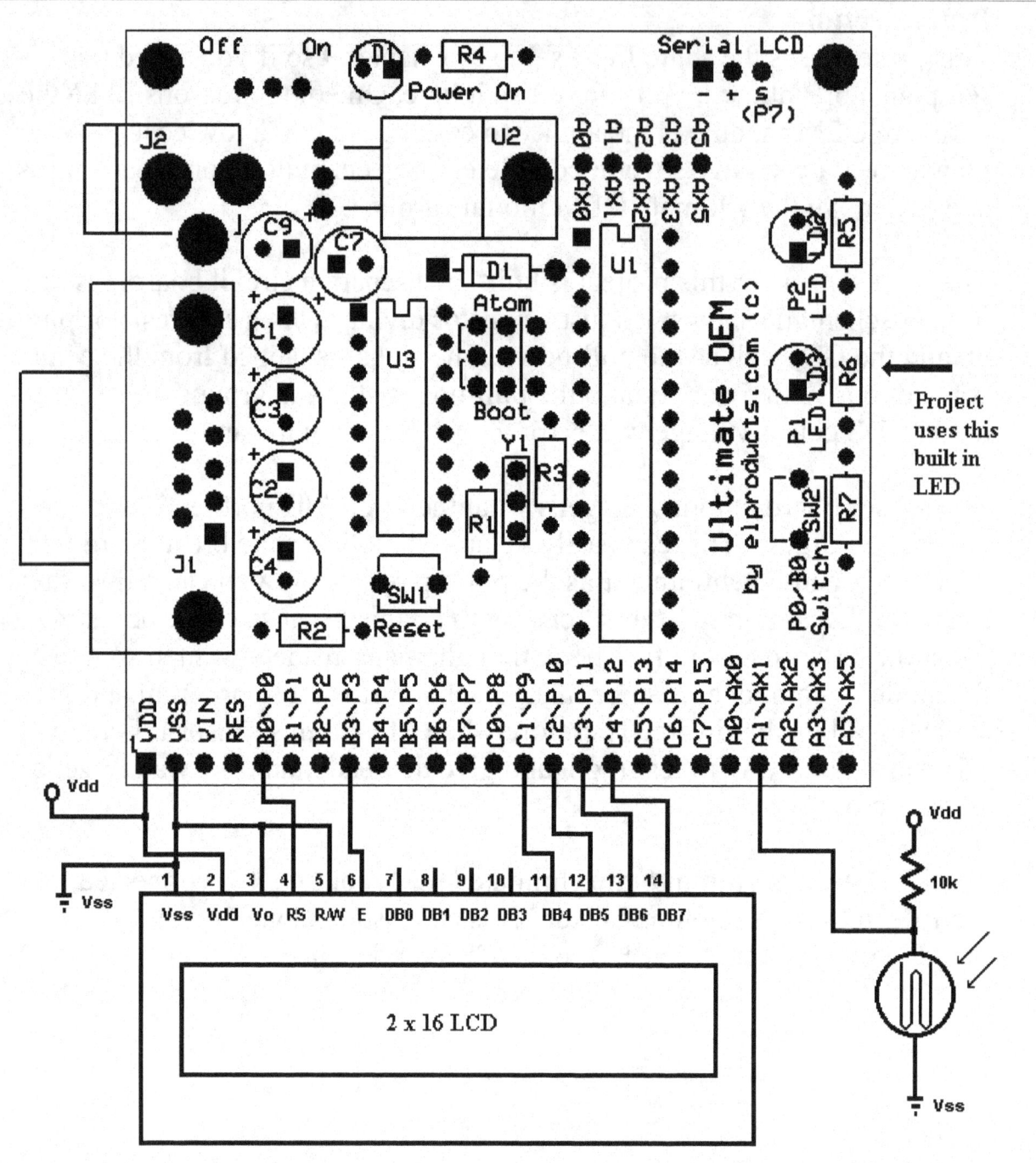

Software

This program steals a lot of the code from the LCD project in Chapter 9. It adds the A/D section to the main loop. This project is a great example of how programming in Basic makes things easier. Being able to read a sensor and display the results in this short list of commands is amazing to me. It even makes a decision to light an LED based on the value measured. That LED could easily be replaced with a relay like the setup used in Chapter 8.

```
epin con 3                  'Nickname for LCD E pin
rspin con 0                 'Nickname for LCD RS pin
sensor con AX1              'Nickname for A/D port
value var word             'Variable for A/D conversion

'**** Setup LCD module for 2x16 display ******
pause 500
lcdwrite rspin\epin,outc,[initlcd1,initlcd2,twoline,scrblk,clear,home]

main
adin sensor,3,ad_ron, value         ' Read Sensor value with A/D
lcdwrite rspin\epin,outc,[clear,home,dec value] ' Display value

pause 100                   ' Delay for smoothness
if value > 900 then         ' Compare size of A/D value
high p1                     ' Value greater than 900 LED on
else                        ' Otherwise do the next commands
low p1                      ' Value lower than 900 LED off
endif                       ' End of the If-Then-Else command
goto main                   ' Loop back and do it again
```

How it works

The beginning of the code establishes the variables and nicknames. The "epin" and "rspin" are for the LCD. The nickname "sensor" is connected to the AX1 A/D port so it's a little easier to understand what pin we are referring to. This is a good habit to get into especially if you have multiple sensors hooked up. Finally a word variable is setup to store the 10 bit A/D value. The Atom micro outputs a 10 bit value so a byte variable just isn't big enough. This will give a range of 0 to 1024 digital so a word is necessary.

```
epin con 3
rspin con 0
sensor con AX1
value var word
```

Next the software delays for 500 milliseconds to let the LCD warm up. Then it sets the LCD display up as a two line display, clears the display and positions the cursor to home in the LCDWRITE command line below.

```
pause 500
lcdwrite rspin\epin,outc,[initlcd1,initlcd2,twoline,scrblk,clear,home]
```

The decision making section of code is located at the label "main". The first line uses the ADIN command to read the sensor level and stores the analog to digital conversion value in the variable "value".

```
adin sensor,3,ad_ron, value
```

The analog to digital converter was discussed in detail in Chapter 2 and I covered some of the ADIN command setup but I cover some of it again here. The command line needs the A/D pin designated, which is the AX1 pin renamed "sensor". After that the number "3" designates to use the internal RC clock for the sampling clock. The "AD_RON" entry, as discussed in Chapter 2, shifts the 10-bit result right in the 16-bit variable area. This means the six most significant bits are zero. Finally the value is stored in the variable "value".

The result is sent to the LCD display as a decimal value. This display will change as the light on the sensor is changed.

```
lcdwrite rspin\epin,outc,[clear,home,dec value]
```

After the ADIN command line the program pauses for 100 milliseconds and then makes a decision to light the LED or not based on the value of the variable "value". The decision is made using an IF-Then-Else command. If the light sensor A/D value is greater than 900 the LED is turned on (high p1). Otherwise the LED is turned off (low p1). A higher A/D value represents a lower amount of light present on the sensor. This means the LED will light when it's closer to dark where the sensor is located.

```
if value > 900 then
high p1
else
low p1
endif
```

After the decision is made, the program jumps back up to the "main" label and takes another reading. The 100-millisecond delay is in the main loop just to make the program seem stable and not jumpy as the light may vary a bit at the sensor.

Next steps

The program can be expanded to control something with the light sensor. The LED could be replaced with a relay to control a higher current light. This setup could possibly be used with a light beam from a laser pointer as the light source shining on the CDS cell to form an invisible line to detect if someone walks by and then count the number of people who walked by in a day.

You don't have to stick with measuring light. The light sensor can be replaced with any resistive type sensor such as a force sensor to measure force not light. This is really a useful project to build from.

Chapter 12 – Making Music

Description

The Atom is not a musical instrument by any means but it can drive a speaker and it does have the ability to play different notes. With that capability this project demonstrates how to use a speaker and the SOUND command to play a crude version of "Mary Had A Little Lamb".

Project setup

The Ultimate OEM module has the speaker connected to the P8 pin of the module thru a 10 uf 25v electrolytic capacitor. The P8 pin is the same as the PIC PortC bit 0 pin for those wanting to know the true PIC connection. In the picture the speaker is actually soldered to a circuit board with the capacitor already in series. I built this board many months ago along with other small breadboard modules to make building projects easier. You can connect the speaker to the breadboard directly and use any 8 ohm speaker as long as it's not too large. Larger speakers tend to require more power which the Atom cannot supply directly.

If you notice, the speaker circuit board also has a potentiometer but it's not wired directly to the speaker as you might think. I put it there hoping to read the potentiometer for volume adjustment but it got too complicated for this project so it just came along for the ride.

The schematic shows the simple connections and demonstrate how easy it is to add sound to any Atom project. The capacitor couples the digital output of the Atom\PIC chip to the analog speaker coil. This allows the speaker to be driven without any additional special electronics such as an amplifier. This also means the speaker will never play anything super loud. It will be loud enough for most of your experiments though.

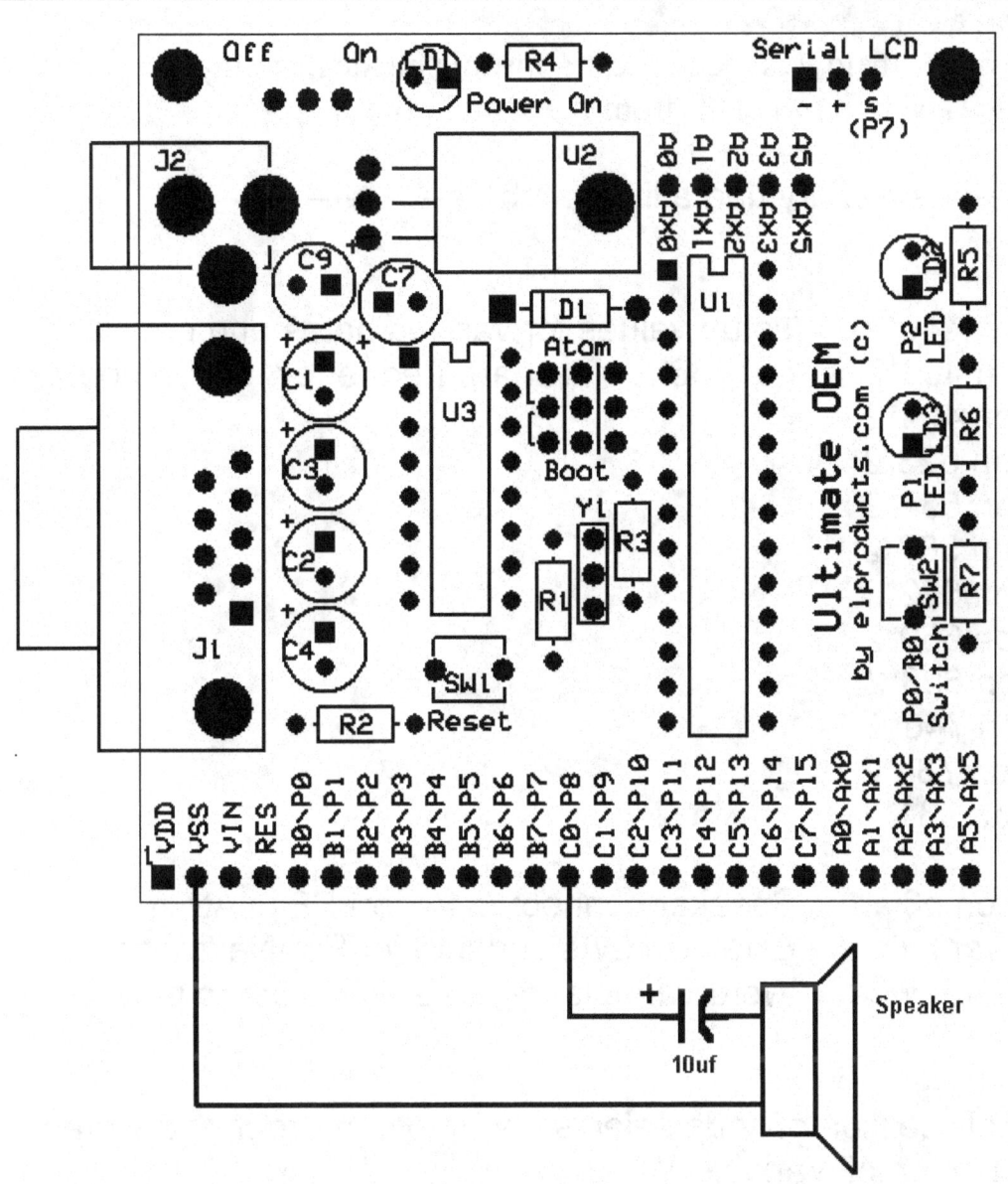

Software

The program below is the orchestra for this musical example. As usual, you can type it in or load it from the sample files.

```
'------------------Mary had a little lamb------------------------


AH con 440*2      'setup values for various notes and octaves
AS con 466*2      'i.e. 466*2 = A sharp frequency, second octave
BH con 494*2
CH con 523*2
CS con 554*2
DH con 587*2
DS con 622*2
EH con 659*2
FH con 698*2
FS con 740*2
GH con 784*2
GS con 831*2


spkr con p8       'Speaker connected to port P8 in Atom
temp var byte     'Establish byte variable for For-Next loop
temp2 var word    'Word variable for storage of note to play.

main
'**** This section of code selects each note in order and stores
'****  it in temp2 variable ***
for temp = 0 to 33

lookup temp,[CS,BH,AH,BH,CS,0,CS,0,CS,BH,0,|
BH,0,BH,0,CS,EH,0,EH,CS,BH,AH,BH,CS,0,CS,0,|
CS,BH,0,BH,CS,BH,AH],temp2
```

```
if temp2 = 0 then skip   'If note value is 0 then play nothing
sound spkr,[500\temp2]        'Send note frequency to speaker


skip:
next                        'Loop back and get next note
pause 5000                  'Pause 5 seconds and play the tune again

goto main                   'Jump to program beginning
```

How it works

The beginning of the program is really the hardest part and that's where the notes are defined. The notes are defined as constants tied to their frequency. Since these are already defined for some musical notes this should make it easier to modify this program to play other songs.

```
AH con 440*2  'setup values for various notes and octaves
AS con 466*2  'i.e. 466*2 = A sharp frequency, second octave
… etc.
```

The speaker connection is defined next as a constant and two variables are established. One is a byte size and the other a word size. The byte variable will be used as the For-Next loop counter. The word size variable is used to store the different notes as they are played.

```
spkr con p8       'Speaker connected to port P8 in Atom
temp var byte     'Establish byte variable for For-Next loop
temp2 var word    'Word variable for storage of note to play.
```

The main loop of code is at the "main" label. It contains a For-Next loop with the SOUND command stuck in the middle. The notes to be played are placed in order within a LOOKUP command table. Between some of the notes a short delay is defined as the value "0".

The LOOKUP command is too long to fit within the page so I used the "|" symbol at the end of each line. The "|" symbol is a code that tells the Atom software compiler the command line continues on the next line. This allows the command line to be very long but fit in a normal printout. This example took two "|" symbols.

```
main
'**** This section of code selects each note in order and stores
'****  it in temp2 variable ***
for temp = 0 to 33

lookup temp,[CS,BH,AH,BH,CS,0,CS,0,CS,BH,0,|
BH,0,BH,0,CS,EH,0,EH,CS,BH,AH,BH,CS,0,CS,0,|
CS,BH,0,BH,CS,BH,AH],temp2
```

The LOOKUP command converts the For-Next loop counter value, in variable "temp", to each unique note and stores it in the "temp2" variable. The "temp2" variable is then used by the SOUND command. If the "temp2" value equals zero then the SOUND command is skipped by jumping to the "skip" label. After the "skip" label the For-Next loop continues with the next value.

```
if temp2 = 0 then skip   'If note value is 0 then play nothing
sound spkr,[500\temp2]       'Send note frequency to speaker
skip:
next                            'Loop back and get next note
```

Finally, after all the 34 notes have been played the program gives the listener a break and delays for 5 seconds. That delay is performed by a simple PAUSE command.

```
pause 5000                'Pause 5 seconds and play the tune again
```

After the delay, the program jumps back to the top using a GOTO command and plays Mary all over again.

Next steps
A simple next step would be to arrange the notes to play a new tune. Another option would be to add more musical frequencies for more complex songs. You could have several songs stored as separate lookup command lines and then select the song via a menu displayed on an LCD and switches to select your choice.

A more useful thought is to give an audible feedback to the user of an Atom based product similar to the way a car chimes when you leave your keys in the ignition. Even feedback when a button is pressed is easy to implement and lets the user know he button press was received.

Chapter 13 – PC to Atom Communication

Description

This project demonstrates how to use the PC to Atom connection for more than just programming. The Atom and PC will communicate through the serial port and use one of the terminal windows, built into the Atom programming software, as the communication screen.

The program will send out a question to the PC that asks which LED to light. That question will show up in the PC terminal window. The user then enters the LED number (0-7) and when the PC "ENTER" key is pressed, the Atom module will receive the value and light the LED connected to that I/O pin and turn off any previously lit LED.

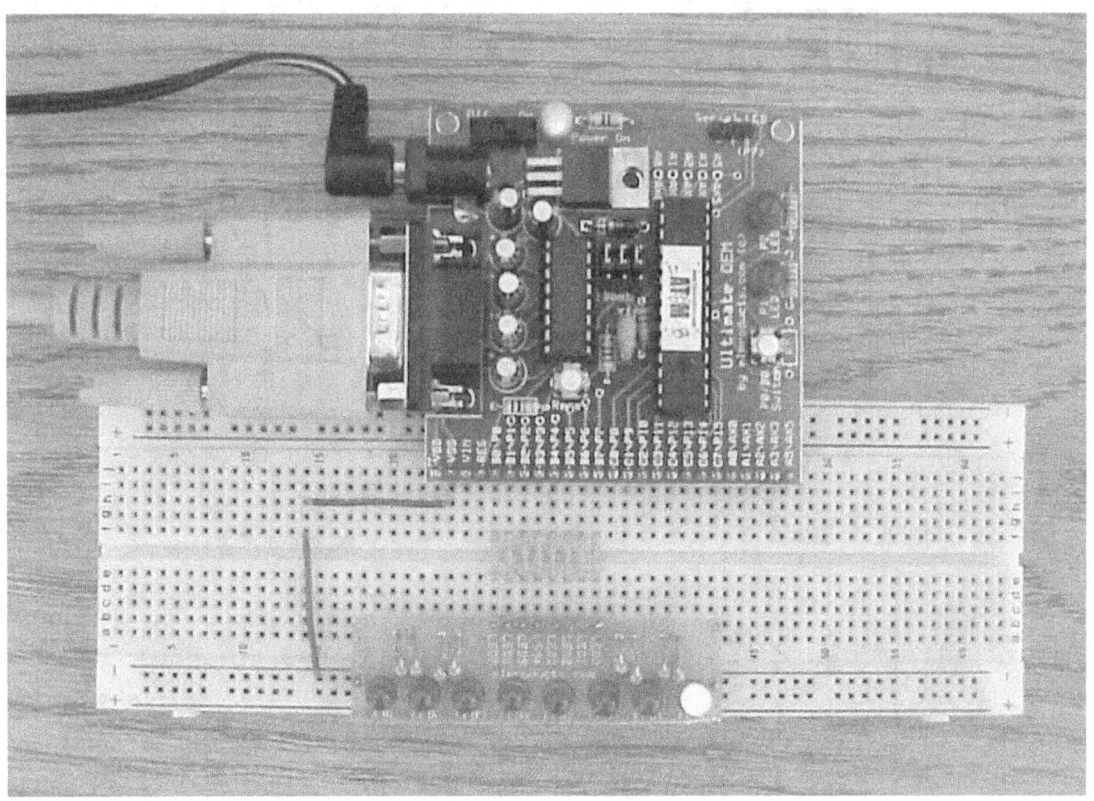

Project setup

As mentioned this project uses the programming port in a dual role. When the program has been downloaded and running in the Atom, the 9-pin header can be used to communicate back to the PC thru the same serial port used to program.

The rest of the setup is similar to the scroll LED's project earlier. Each LED is tied to a port B pin or P0-P7. The LED's get power from the port and are commonly grounded thru the Vss pin of the Ultimate OEM Atom module. Each LED is current limited by a 220 ohm ¼ w or larger resistor.

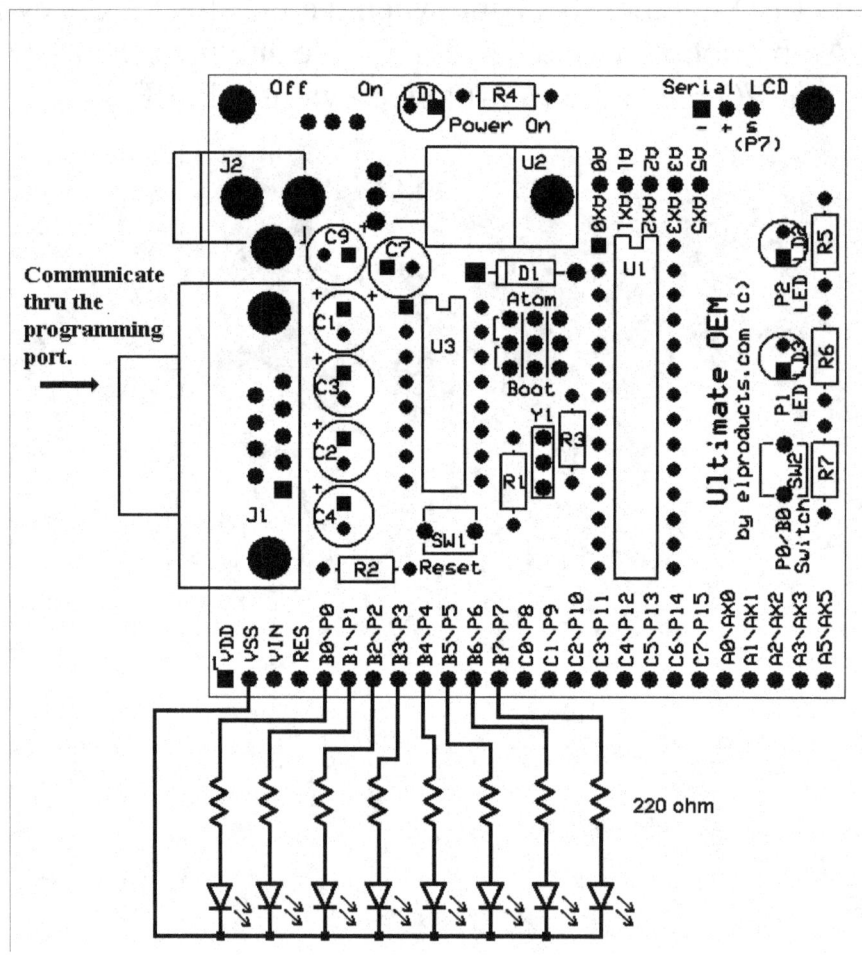

The picture below is a screen shot of the Atom terminal window. You get there by clicking on the tabs at the bottom of the screen. The selection boxes at the top of the terminal window are the communication parameters. The only parameter you have to change from the default level is the baud rate. This program communicates at 9600 baud.

After the program is running in the Ultimate OEM , bring up the terminal window, set the baud rate to 9600 and then click on the "connect" button. It should then ask you which LED to light. If it doesn't, press the Ultimate OEM reset button to restart the program in the Atom.

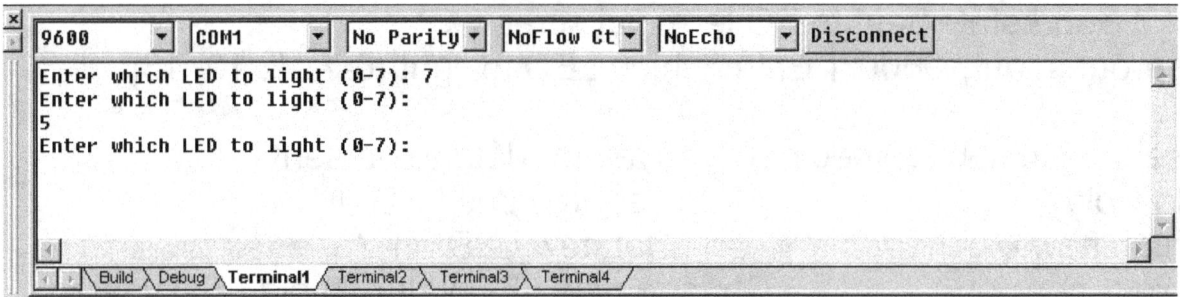

To make a selection, click in the description window and type a number from 0 to 7 and then hit enter. The LED you typed should now be lit on the setup.

Software

The program below is quite simple but a great start to make menus or control screens for your Atom project. As usual, you can type it in or load it from the sample files.

```
old var byte            ' Variable to store the previous selection
new var byte            ' Variable to store the new selection


main


'*** Send statement to PC terminal screen ***
serout s_out, i9600, ["Enter which LED to light (0-7): ", 10, 13]


serin s_in, i9600, [dec new]  ' Wait for data to be sent
low old                       ' Set previous LED off
high new                      ' Set new LED on
old = new                ' Change new LED value to old LED value
pause 100                ' Pause 100 milliseconds to see the LED


goto main                ' Loop back and do it again
```

How it works

The best part about this project is you don't need to know all the technical details about PC serial communication to make this work. You just have to know the baud rate used and how to use a terminal window as described above.

The program starts off by setting up a couple of byte variables "old" and "new". These are used to store the previous LED lit and the new LED to light.

old var byte
new var byte

At the "main" label, the program sends the LED question to the PC terminal window. It does this with the SEROUT command. The pin label used to communicate is the S_OUT label, which is the same pin already connected to the level inverter circuitry for programming the Atom. Every Atom uses that same label for that pin. The Atom programming software communicates at a much higher baud rate and in a different protocol so using this communication method does not reprogram the Atom.

```
main
'*** Send statement to PC terminal screen ***
serout s_out, i9600, ["Enter which LED to light (0-7): ", 10, 13]
```

After the S_OUT pin is set the baud rate follows. The "i9600" is used as the communication rate. The "i" part indicates inverted mode because the signal is coming from the PIC and then inverted by the level shifter circuitry in the Atom module.

After the baud rate is set then the LED question is placed inside the brackets between quotes. Everything between quotes is sent as ASCII data. In other words it's sent as letters to be displayed not numerical data. In this case "Enter which LED to light (0-7):" is sent. The 10 and 13 that follow are the ASCII value codes for line feed and carriage return. This positions the data sent on a line of it's own and not a continuation of the previous line.

After sending the question, the Ultimate OEM just waits at the next line for a response. The SERIN command is used for this. The communication pin is the S_IN pin that is already connected to the level inverter circuitry in the Ultimate OEM. It's the "input" companion to the S_OUT label. The "i9600" baud rate is used again to keep the communication the same.

```
' *** Receive from PC which LED to light ***
serin s_in, i9600, [dec new]
```

When the LED value is received, the value is stored in the byte variable "new" as seen inside the brackets. In front of the "new" variable name is the modifier "DEC". "DEC" is short for decimal and converts the ASCII byte value received into a decimal number. You see, all communication between the PC and Atom module is by ASCII characters. If the user entered "7" for the 7th LED then the ASCII character value for "7" (hex $25) is sent not the decimal number 7. By putting the "DEC" in front of the variable, the SERIN command converts the ASCII value to the decimal value equivalent.

Finally the program turns off the previous lit LED and turns on the new LED. This is done with simple high and low commands. The program then stores the new LED as the old LED before jumping back up to the top of the program to run it all again.

```
low old      ' Turn off previous LED
high new     ' Turn on present LED
old = new    ' Make present LED previous LED
pause 100    ' Delay 100 milliseconds
goto main
```

Next steps

This whole project can be expanded to have the terminal window control anything connected to the Atom module I/O. If you are good at PC programming you can use Visual Basic or similar PC programming tools to develop a custom interface window that controls an Atom module using this same SERIN/SEROUT method.

You can use this project with many of the earlier projects to control an LCD or Relay direct from the PC screen. Industrial controls use this method often. A PC running Labview software will have multiple buttons and slide switches to adjust. Labview then sends out the data through the serial port. The Atom can be the receiving control module that actually turns things on an off. It's just a matter of how far you want to expand the use of the Atom.

Chapter 14 – Internal EEPROM

Description

Inside the Atom micro is 256 bytes of non-volatile Electrically Erasable Programmable Read Only Memory (EEPROM). This is memory that will stay in-tact even when the power is removed. This project shows how to access and setup that memory. We store some simple phrases that state which switch was pressed. This could launch off the previous project and have different menu text or port control data stored in EEPROM and controlled thru a PC display.

This project has four switches tied to I/O ports. When one of the switches is pressed, a statement about which switch was pressed is pulled from EEPROM and displayed on an LCD screen. This is simple but effective at demonstrating how to use EEPROM.

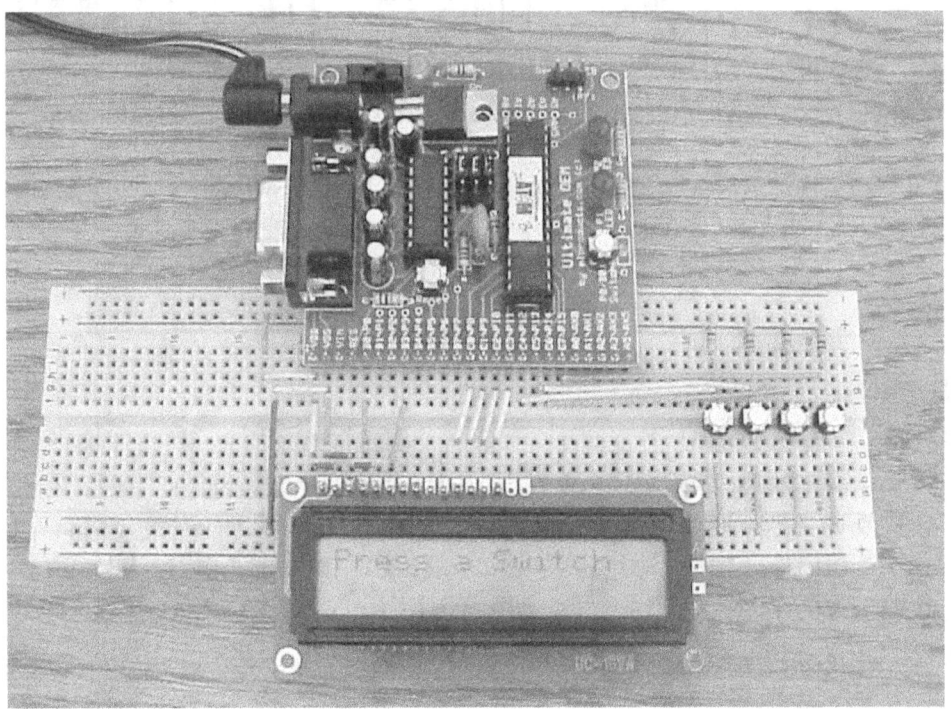

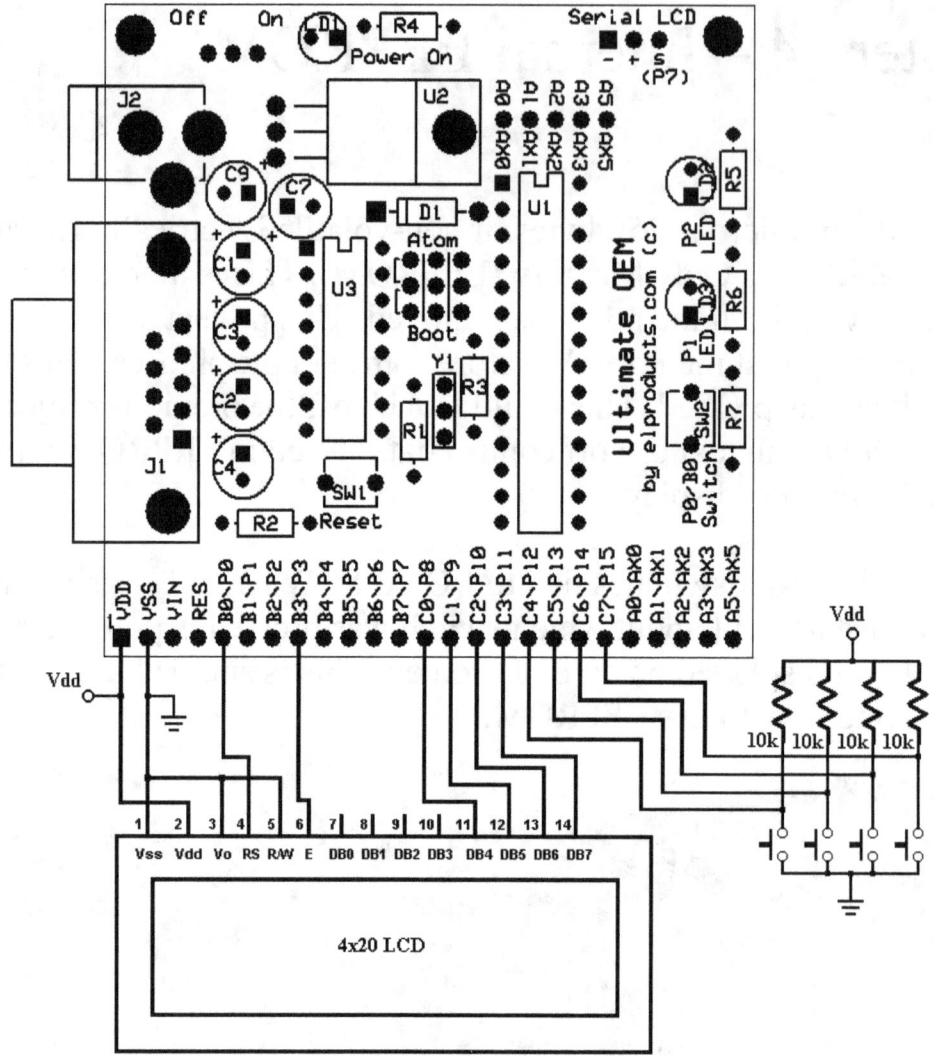

Hardware setup

The hardware is very similar to other projects. Switches are connected to the P12 thru P15 pins. Each has a 10k pull-up resistor to keep the port high when not in use. We cannot use the internal pull-ups because they are only available on the Port B pins, which are P0 thru P7. The LCD is wired the same as previous projects. If you saved that setup you can easily just add the switches.

Software

The program below is setup similar to other projects in this book, quite simple but a great start to expand on. You can type it in or load it in from the sample files.

```
sw1 var in12            'Switch 1 connected to P12
sw2 var in13            'Switch 2 connected to P13
sw3 var in14            'Switch 3 connected to P14
sw4 var in15            'Switch 4 connected to P15
x var byte              'General Purpose Variable
addr var byte           'Stores EEPROM address
char var byte           'Stores EEPROM value
rspin con 0             'LCD RS Pin
epin con 3              'LCD E Pin


'*** Setup initial EEPROM data ***
data @0,"switch1",@10,"switch2",@20,"switch3",@30,"switch4"

'*** Setup LCD display ***
pause 500
lcdwrite rspin\epin,outc,[initlcd1,initlcd2,twoline,scrblk,clear,home]

'*** Main program loop ***
main

'*** Display Project title on LCD for 1 second***
lcdwrite rspin\epin,outc,[clear,home,"Chapter 14",scrram+$40,"EEPROM"]
pause 1000

'*** Display request for user to press a switch ***
again
lcdwrite rspin\epin,outc,[clear,home,"Press a Switch"]
```

```
pause 100

if sw1 = 0 then          'Test if Switch 1 is pressed
hold1:
if sw1 = 0 then hold1    'Switch pressed, wait for it to release
addr = 0                 'Set EEPROM address value to zero
goto display             'Jump to read EEPROM routine
endif                    'End this If-Then command


if sw2 = 0 then          'Test if Switch 2 is pressed
hold2:
if sw2 = 0 then hold2    'Switch pressed, wait for it to release
addr = 10                'Set EEPROM address value to ten
goto display             'Jump to read EEPROM routine
endif                    'End this If-Then command


if sw3 = 0 then          'Test if switch 3 is pressed
hold3:
if sw3 = 0 then hold3    'Switch pressed, wait for it to release
addr = 20                'Set EEPROM address value to twenty
goto display             'Jump to read EEPROM routine
endif                    'End this If-Then command


if sw4 = 0 then          'Test if switch 4 is pressed
hold4:
if sw4 = 0 then hold4    'Switch pressed, wait for it to release
addr = 30                'Set EEPROM address value to thirty
goto display             'Jump to read EEPROM routine
endif                    'End this If-Then command


goto again               'Jump back to "again" label to re-read
switches
```

```
'*** Read EEPROM and Display on LCD routine ***
display
lcdwrite rspin\epin,outc,[clear,home]      'Clear LCD screen
for x = 1 to 7                             'Loop through 7 characters
read addr, char                            'Read character from EEPROM
addr = addr + 1                            'Move to next character address
lcdwrite rspin\epin,outc,[char]            'Display character on LCD
next                                       'Get next Character from EEPROM
pause 1000                                 'Wait 1 second before returning
goto again          'Jump back to "again" label to re-read switches

goto main          'If all else fails go back to beginning
```

How it works

First we establish the variables and create names for the switch inputs.

```
sw1 var in12          'Switch 1 connected to P12
sw2 var in13          'Switch 2 connected to P13
sw3 var in14          'Switch 3 connected to P14
sw4 var in15          'Switch 4 connected to P15
```

Next we create a general-purpose variable and two variables to store the EEPROM memory control bytes. We also define the LCD connections.

```
x var byte            'General Purpose Variable
addr var byte         'Stores EEPROM address
char var byte         'Stores EEPROM value
rspin con 0           'LCD RS Pin
epin con 3            'LCD E Pin
```

Now storing the switch statements initializes the EEPROM. We use the DATA command to do this. Each string is in quotes and is stored as ASCII bytes. Listed before each string is the memory location where the strings are stored.

```
'*** Setup initial EEPROM data ***
data @0,"switch1",@10,"switch2",@20,"switch3",@30,"switch4"
```

The first is at the top or location zero. The next is placed at location 10 thus leaving 10 bytes (location 0 thru location 9) for the first phrase. The third phrase is at location 20 and the fourth at location 30. You can place these anywhere you want just doesn't exceed the 0 - 255 limit that exists because the EEPROM only has 256 bytes of space.

Next we setup the LCD the same way previous programs did.

```
'*** Setup LCD display ***
pause 500
lcdwrite rspin\epin,outc,[initlcd1,initlcd2,twoline,scrblk,clear,home]
```

Now we enter the main loop of code. We start by displaying a project title on the LCD to prove that everything is working.

```
'*** Main program loop ***
main
```

```
'*** Display Project title on LCD for 1 second***
lcdwrite rspin\epin,outc,[clear,home,"Chapter 14",scrram+$40,"EEPROM"]
pause 1000
```

Next, the program displays a question asking the user to press a switch.

```
'*** Display request for user to press a switch ***
```

```
again
lcdwrite rspin\epin,outc,[clear,home,"Press a Switch"]
pause 100
```

Now we enter the section that checks the switches. This is very similar to how the Chapter 7 project read switches with the IF-THEN command. I'll just describe one switch read since it just repeats for each switch.

It starts by checking if the pin connected to the switch is low. If it is, the switch has been pressed so then next command at the label "hold1" is processed. This command line does the same thing, looking for the switch pin to be low but it sends the program back to the "hold1" label until the switch is released. This puts the program in a continuous loop until the switch is released.

```
if sw1 = 0 then          'Test if Switch 1 is pressed
hold1:
if sw1 = 0 then hold1     'Switch pressed, wait for it to release
```

This method is kind of a crude de-bounce routine but works great for applications where the switch could get multiple hits in a short time. After the switch is released, the program then sets the address variable to zero which is where the "switch1" phrase is stored. The program then jumps to the "display" label to retrieve the proper phrase from EEPROM and displays it on the LCD. We finish this command with the ENDIF command to end the IF-Then loop.

```
addr = 0                 'Set EEPROM address value to zero
goto display             'Jump to read EEPROM routine
endif                    'End this If-Then command
```

All this repeats for the other three switches and then a GOTO command returns the loop back to the top to read the switches again. If no switch was pressed, each If-Then loop would be quick and the GOTO command will send it back to the top to check again thus creating a continuous check of the switches.

```
goto again          'Jump back to "again" label to re-read switches
```

Here is where the EEPROM is accessed. First the LCD screen is cleared.

```
'*** Read EEPROM and Display on LCD routine ***
display
lcdwrite rspin\epin,outc,[clear,home]        'Clear LCD screen
```

Then the program uses the general variable "x" to count thru seven characters, thus pulling one letter at a time from EEPROM. The READ command is used to pull a character byte from EEPROM. The address, setup by the "addr" variable in the switch routine, tells the READ command where to look in EEPROM for the data. We know each phrase contained seven characters so we just loop through this seven times.

```
for x = 1 to 7                   'Loop through 7 characters
read addr, char                  'Read character from EEPROM
```

Each time thru the loop the "addr" variable is incremented by one to get the next character. Each character retrieved from EEPROM is sent to the LCD with the LCDWRITE command line. This continues until all seven characters have been retrieved.

```
addr = addr + 1                  'Move to next character address
lcdwrite rspin\epin,outc,[char]  'Display character on LCD
```

The program then delays for 1 second before checking the switches by using the GOTO command to jump back to the label "again".

```
pause 1000          'Wait 1 second before returning
goto again          'Jump back to "again" label to re-read switches
```

Just in case the program some how slips through everything, we add a GOTO command to put the program back to the "main" label. The program should actually never reach this point.

```
goto main           'If all else fails go back to beginning
```

Next steps

EEPROM has multiple uses for storing information. This is just one simple example of how to store text in EEPROM. This could be expanded to contain multiple menus that are displayed on an LCD depending on which switch is pressed. Numerical data can be stored in EEPROM and then retrieved and used in a math equation based on a switch combination. That's all a calculator does when you store a value in memory and then retrieve it later for another calculation. The main difference is the calculator stores it in RAM not EEPROM so the data is lost when you turn the calculator off.

EEPROM can be very handy to store numbers and text together and keep it for a long time. What I did not demonstrate is the WRITE command, which stores data in EEPROM rather than read data from EEPROM. This program could be easily modified to have it store data when one button is pushed and retrieve it when another button is pushed. Now that I think of it I should have done that here. Well, I guess I leave that up to you to try. Let's move to the next project.

Chapter 15 – Driving LEDs with Max7219

Description

This project demonstrates how to control a Maxim Max7219 display driver using the Atom. The project is quite simple and I suggest you read the max7219 data sheet that you can download from maxim-ic.com. It operates similar to a simple shift register but has many special features to make driving LED displays quite easy. The max7219 can drive up to eight displays but for this project I'll just drive three and have the project count up from 000 to 999.

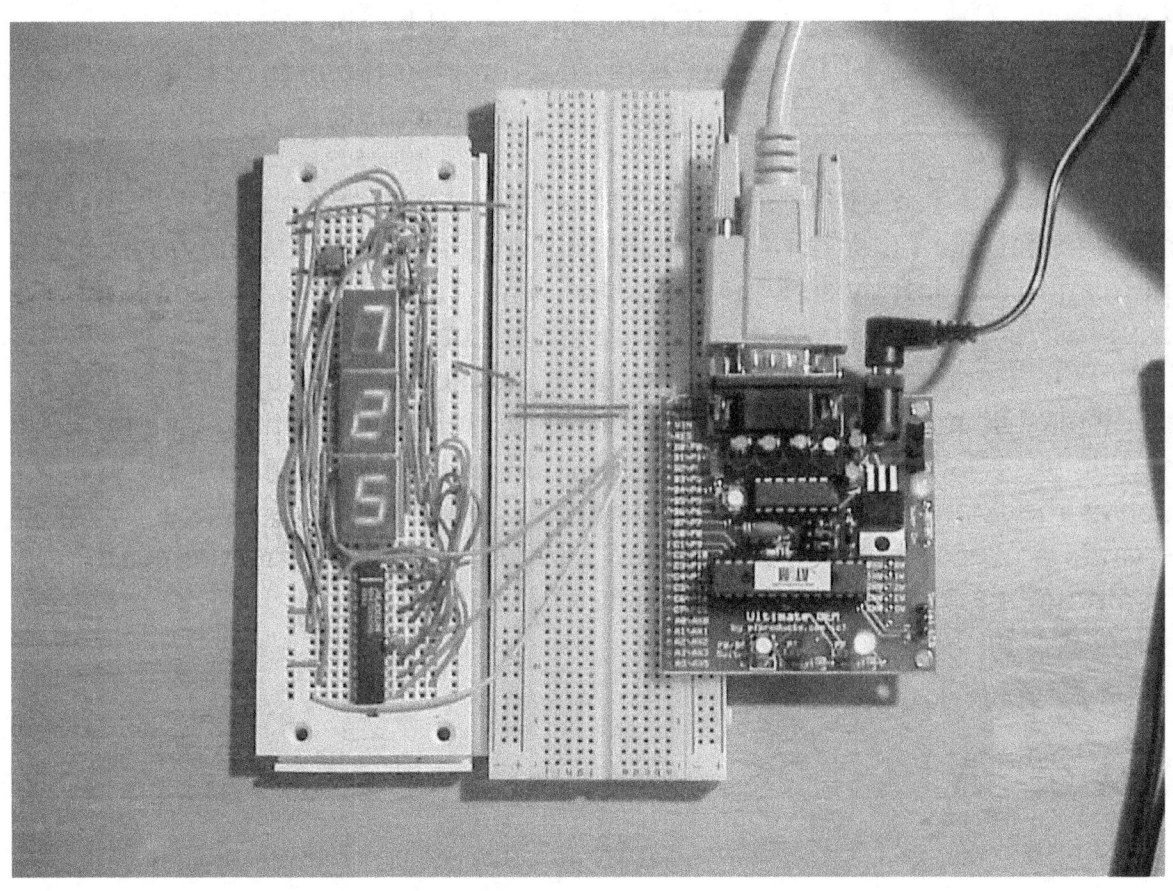

Project setup

The max7219 and LED's are built on a separate breadboard. This is for two reasons; 1) it was easier to fit and it can easily be added to any project that I want to add the display to and 2) I had built this several years ago for a different project so I was too lazy to rewire it on the same board as the Atom.

This is one of the more complex schematics in this book but it's really not that tough to build. The max7219 drives common cathode displays which just means every individual LED in the display has their cathodes all connected at one pin. The LED's are multiplex driven by the max7219 which means the displays all share the same connections to their anodes and then each separate cathode is turned on or off by the max7219. Therefore, only one LED display is on at a time but the max7219 drives them in succession so fast the human eye picks them up as all lit together.

Now you could use LED's that mount horizontal on the breadboard rather than vertical like I did, but I wanted it to look like this in the previous application so I left those LED's in place. The Atom connections are quite easy. Just a clock, data and enable line are all that is needed. Using the SHIFTOUT command makes this setup easy to control. I supply power to the display board from the Ultimate OEM module Vdd output since this module has a larger 5-volt regulator. If you use a different Atom module, such as the Atom 24 pin, you might want to power the LED's via a separate power source since the Atom 24 regulator cannot handle nearly the amount of current that the Ultimate OEM can.

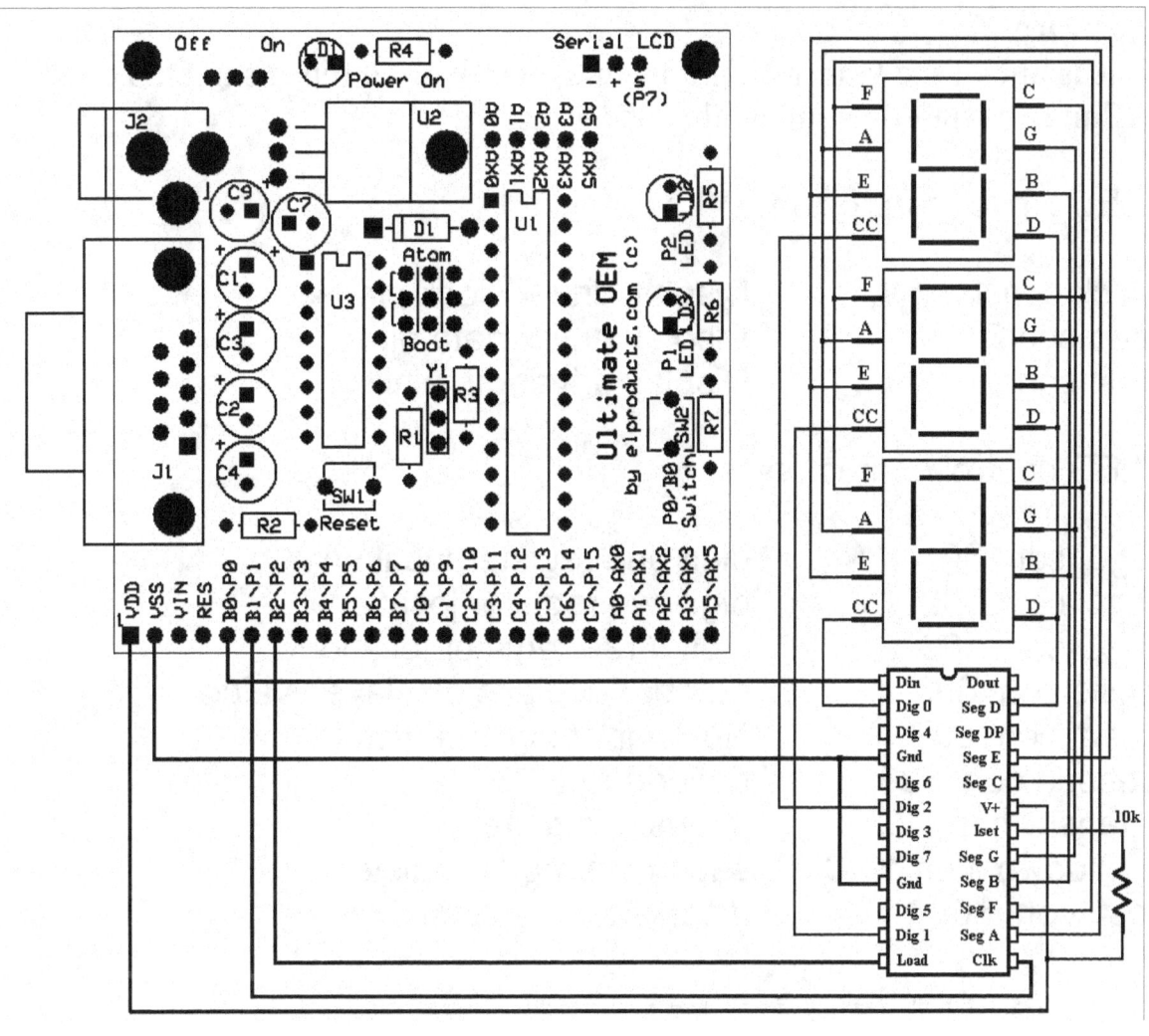

Software

This is one of the longer programs in the book but it repeats a lot. You'll definitely want to download this one.

```
' LED Driver control pins
'

clock con 1              ' Display driver clock pin
load con 2               ' Display driver data pin
dpin con 0               ' Display driver din pin

' LED control characters
'

dig0  con  $01           ' digit one register location
dig1  con  $02           ' digit two register location
dig2  con  $03           ' digit three register location
digits con  $02          ' number of digits displayed value
scan  con  $0B           ' Number of digits register
decode con  $09          ' decode register
intensity con  $0A       ' intensity register
shutdown  con  $0C       ' shutdown mode register
test  con  $0F           ' display test register

' -----[ Variables ]----------------------------------------
'

x var byte               ' General purpose variable
y var word               ' General purpose variable
dindata var word         ' Display data for din input
register var byte        ' Register locator
databyte var byte        ' Data information
one var byte             ' 1's digit
ten var byte             ' 10's digit
hund var byte            ' 100's digit
```

```
' -----[ Initialization ]------------------------------------
'
Init:
        PortB = $00              ' All Port B outputs off to start
        TrisB = %11111000        ' Pins 1,2,3 outputs
        PAUSE 10                 ' Pause for driver setup

' Initialize the LED Driver
'

I_Driver:
        databyte = 1             'all segments on
        register = test         ' 500 msec
        gosub datasend

        databyte = $FF           'Intensity Adjust to max
        register = intensity
        gosub datasend

        databyte = $FF           'decode code b for all digits
        register = decode
        gosub datasend

        databyte = digits        ' 3 digits displayed
        register = scan
        gosub datasend

        databyte = 1             'Start Normal Operation
        register = shutdown
        gosub datasend

        databyte = 0             'Initialize zero digit to 0
        register = 0
```

```
        gosub datasend

        register = 1              'Initialize ones digit to 0
        gosub datasend

        register = 2              'Initialize tens digit to 0
        gosub datasend

        databyte = 0             'All LEDs test mode off
        register = test
        gosub datasend

' -----[ Main Code ]-------------------------------------
'
'Display initial screen
'

reset:
        one = 0                  ' Clear one's digit byte
        ten = 0                  ' Clear ten's digit byte
        hund = 0                 ' Clear hundred's digit byte
        goto display             ' Send these values to display

start:
one = one + 1                    ' Increment one's byte

if one < 10 then display         ' Test if one's digit is less than 10
one = 0                          ' If less than 10 send to display
ten = ten + 1                    ' If not, reset one to zero and
                                 ' increment the ten's byte

if ten < 10 then display         ' Test if ten's digit is less than 10
ten = 0                          ' If less than 10 send to display
hund = hund + 1                  ' If not, reset ten to zero and
```

```
                                        ' increment the hundred's byte
if hund < 10 then display               ' Test if hund's digit is less than 10
goto reset                              ' If less than 10 send to display
                                        ' If not, 999 is achieved, go to reset


'*** Send display data to the Max7219 ***

display:
        databyte = one          ' Send One's digit value
        register = dig0         ' to the first LED
        gosub datasend

        databyte = ten          ' Send 10's digit value
        register = dig1         ' to the second LED
        gosub datasend

        databyte = hund         ' Send 100's digit value
        register = dig2         ' to the third LED
        gosub datasend

        databyte = $ff          ' Send max bright level
        register = intensity    ' to all LEDs
        gosub datasend

        pause 1000                      ' Delay 1 second
        goto start              ' Jump back to beginning


'*** Send command byte to LED driver Subroutine *** '

datasend:
low load                        ' Enable load mode on Max7219
dindata.lowbyte = databyte      ' Combine Databyte and Register
```

```
dindata.highbyte = register    ' byte into one word

shiftout dpin, clock, 0,[dindata]    ' Send word to Max7219
high load                             ' Disable load mode on Max7219

return                                ' Return from subroutine
```

How it works

This project is one of the more useful ones in this book. One of the more common questions I get email about is how to drive LED displays with the Max7219. It's not much different than driving a shift register. We start the program by establishing the Max7219 control pin connections using the CON directive.

```
' LED Driver control pins
'

clock con 1              ' Display driver clock pin
load con 2               ' Display driver data pin
dpin con 0               ' Display driver din pin
```

The Max7219 has special control codes it uses to set it up. To make it easier to understand the program we create nicknames for the control codes using the CON directive. The control codes come from the Max7219 data sheet which I've included on the book CD.

```
' LED control characters
'

dig0  con  $01           ' digit one register location
dig1  con  $02           ' digit two register location
dig2  con  $03           ' digit three register location
digits con  $02          ' number of digits displayed value
```

```
scan  con  $0B          ' Number of digits register
decode con  $09         ' decode register
intensity con  $0A      ' intensity register
shutdown  con  $0C      ' shutdown mode register
test  con  $0F          ' display test register
```

The variables are setup to hold the values that are sent to the Max7219 plus some general purpose variables to be used in the FOR-NEXT loops. The Max7219 actually receives the data to be displayed as a word. The data to be displayed is 8-bits of it and the register of where it is stored within the Max7219, is the other 8-bits. To make this easier to remember, variables are setup to hold this information in the software. Also included are byte variables to store the 1's, 10's and 100's digit value.

```
' -----[ Variables ]-----------------------------------
'

x var byte              ' General purpose variable
y var word              ' General purpose variable
dindata var word        ' Display data for din input
register var byte       ' Register locator
databyte var byte       ' Data information
one var byte            ' 1's digit
ten var byte            ' 10's digit
hund var byte           ' 100's digit
```

Next we have to get the I/O setup to be outputs. I write directly to the PortB control registers to do this. First I clear port B which controls the Atom pins P0 thru P7. Then I make P0,P1 and P2 outputs by setting the TRISB register to binary %11111000. The last three bits are "0" which make those pins outputs. The "1's" make pins P7 thru P3 inputs. We pause briefly to let the Max7219 settle. This can probably be skipped but I like to let the hardware settle before sending a bunch of commands.

```
' -----[ Initialization ]----------------------------------
'
Init:
        PortB = $00          ' All Port B outputs off to start
        TrisB = %11111000    ' Pins 1,2,3 outputs
        PAUSE 10             ' Pause for driver setup
```

Now that all the variables, nicknames and I/O are setup, we create the code to properly setup the Max7219 to do what we want. For this project we are only driving three LED display digits rather than the eight the Max7219 is capable of. The first set-up item is the display test mode.

```
' Initialize the LED Driver
'


I_Driver:
        databyte = 1              'all segments on
        register = test          ' 500 msec
        gosub datasend
```

The code above sets the variable data byte to the value %00000001 binary. Then it makes the register variable equal to the nickname "test" which was set equal to $0F earlier with the CON directive earlier. Therefore we established the data and the location of where we want to send this control information in the Max7219. This location and data put the Max7219 in test mode which lights all the LEDs on the display and it stays that way until we send that same location a "0" byte value. We just set the value here though and then we jump to the subroutine "datasend" using the GOSUB command.

'*** Send command byte to LED driver Subroutine *** '

```
datasend:
low load                          ' Enable load mode on Max7219
dindata.lowbyte = databyte    ' Combine Databyte and Register
dindata.highbyte = register   ' byte into one word

shiftout dpin, clock, 0,[dindata]        ' Send word to Max7219
high load                                 ' Disable load mode on Max7219

return                                          ' Return from subroutine
```

The "datasend" subroutine is above. I put this into a subroutine because this section of code will get used often. Every time we want to send information to the Max7219, this subroutine will be used. All it needs is the databyte variable value and the register variable value. The subroutine first sets the "load" pin (P2) to a low level using the LOW command. This puts the Max7219 into "receive new data" mode. Then the subroutine combines the databyte and the register byte into the word size variable "dindata". This is done separately by using the modifiers ".lowbyte" and ".highbyte". After these two steps, dindata contains the data to send as the lower 8-bits and the location to put it, in the upper 8-bits.

The subroutine now sends the dindata word to the Max7219 using the SHIFTOUT command. The data pin connection and clock pin connection use the nicknames we made earlier with the CON directive. Also in the SHIFTOUT command line is the number "0". That "0" is the mode selection which represents MSBPRE or Most Significant Bit. In this mode the most significant bit is setup up prior to sending the clock pulse. This is the way the Max7219 wants to receive its data. After the mode value, the "dindata" word is set in the brackets indicating what variable has the data to send. The SHIFTOUT command takes care of sending the clock pulses for each bit and the Max7219 will have the information after the SHIFTOUT command is finished.

After the SHIFTOUT command, we set the load pin back to a high state to set the Max7219 back to operation mode and then the RETURN command is issued to jump the program back to the command after the GOSUB command that sent it to the subroutine.

The next few sections operate the same way as above but send the Max7219 different setup data. Let's go thru those briefly so you understand how the software sets up the Max7219.

```
databyte = $FF        'Intensity Adjust to max
register = intensity
gosub datasend
```

The code above sets the intensity of the LEDs to maximum brightness.

```
databyte = $FF        'decode code b for all digits
register = decode
gosub datasend
```

The Section above establishes how the Max7219 reads the data it receives. It puts it in Decode B mode which is BCD or Binary Coded Decimal mode. This means it really only looks at the lower 4 bits of the data byte and lights the proper segments to display the decimal value of the byte. For example, if %00001001 is sent then the Max7219 would light the segments that form the number "9".

```
databyte = digits   ' 3 digits displayed
register = scan
gosub datasend
```

The section of code above establishes the number of digits to drive. By setting the "databyte" variable to the value of the "digits" nickname (which is 3), then the software is telling the Max7219 to display only 3 digits.

```
databyte = 1                    'Start Normal Operation
register = shutdown
gosub datasend
```

The section above puts the Max7219 in run mode (1) versus shutdown low power mode (0) so it is ready to display the numbers we send it.

```
databyte = 0
register = 0
gosub datasend

register = 1
gosub datasend

register = 2
gosub datasend
```

These next three sections all preset the 1's, 10's and 100's digit to zero by sending the "databyte" value of zero to each of their respective locations which we made nicknames for at the beginning.

Finally, we turn off the test mode by sending a "0" value to that register. This puts the display in normal mode and at that point will display the "000" value that we preset the display values to.

```
databyte = 0
register = test
gosub datasend
```

Now we enter the main section of code where we create the counter values to be displayed on the LEDs.

```
' -----[ Main Code ]----------------------------------------
'
'Display initial screen
'

reset:
one = 0                 ' Clear one's digit byte
ten = 0                 ' Clear ten's digit byte
hund = 0                ' Clear hundred's digit byte
goto display            ' Send these values to display
```

First the counter variables "one", "ten" and "hund" are preset to zero. The program uses the GOTO command to jump to the "display" label.

```
'*** Send display data to the Max7219 ***

display:
        databyte = one          ' Send One's digit value
        register = dig0         ' to the first LED
        gosub datasend

        databyte = ten          ' Send 10's digit value
        register = dig1         ' to the second LED
        gosub datasend

        databyte = hund         ' Send 100's digit value
        register = dig2         ' to the third LED
        gosub datasend

        databyte = $ff          ' Send max bright level
        register = intensity    ' to all LEDs
```

```
gosub datasend

pause 1000                    ' Delay 1 second
goto start                    ' Jump back to beginning
```

The software at the "display" label is very similar to the setup software because it loads the "databyte" variable and the "register" byte variable with the values to be displayed based on the values of the "one", "ten" and "hund" variables. The program uses the same subroutine "datasend" to send the new values to be displayed. Therefore by modifying the variables "one", "ten" and "hund" we can make the display count from 000 to 999 and then reset back to 000. We do that below.

```
start:
one = one + 1                 ' Increment one's byte

if one < 10 then display      ' Test if one's digit is less than 10
one = 0                       ' If less than 10 send to display
ten = ten + 1                 ' If not, reset one to zero and
                              ' increment the ten's byte

if ten < 10 then display      ' Test if ten's digit is less than 10
ten = 0                       ' If less than 10 send to display
hund = hund + 1               ' If not, reset ten to zero and
                              ' increment the hundred's byte

if hund < 10 then display     ' Test if hund's digit is less than 10
goto reset                    ' If less than 10 send to display
                              ' If not, 999 is achieved, go to reset
```

This is a very simple way to count in Atom Basic. We first increment the "one" variable by adding "1" to it. Then we test it to see if it's less than 10. If it is we just jump to the display label, which just updates the display. If instead the value of "one" is not less than 10 then it resets "one" to zero and increments the "ten" variable. Then we test to see if the "ten" variable has rolled over to 10 so we can increment the "hund" variable. If the "hund" variable is greater than 10, we know we are at 1000 so jump to reset which puts the values of "one", "ten" and "hund" back to zero and it starts counting all over again. The program is really not that complex but is very useful.

Next steps

You can see that broken down this long program is hopefully very easy to understand. Driving a Max7219 offers a lot of possibilities for future projects you might want to do with the Atom module. Because the Max7219 can drive more than three LED displays you could easily modify the hardware to add another LED and then modify the software to make it drive four or more LED displays. You could also combine this with the switch project and make the count increase when a switch is pressed.

For some reason, even though LCD's are easier to drive and, as I showed you in Chapter 10, can be made to create large digits, many people still want to use LED displays in projects. Using the Max7219 makes the hardware easy and now you have the basic software for driving it. This should be a handy project for you in the future.

Chapter 16 – Using Timer 1 Interrupt

Description

The Microchip PIC within the Atom has three timer modules, two 8-bit wide and one 16-bit wide. This project demonstrates how to use the 16-bit wide timer as an accurate time base. I didn't want to complicate this project with complicated hardware so I decided to go back to where we started and flash an LED. Instead of relying on the pause command, we use the accuracy of TIMER1. You might question why using PAUSE isn't accurate enough, and I don't want that implied. The PAUSE command is actually quite accurate based on some oscilloscope measurements I've taken in previous projects using PAUSE. I just wanted the hardware to be simple so we could focus on the Timer 1 use.

The 16-bit Timer1 counts from 0 to 65535 incrementing once on every internal clock pulse and then overflows at back to 0 after 65535. When it overflows, it also sets the TIOF bit on the INTCON register of the PIC. This also will trigger an interrupt if you have interrupts turned on in software. I decided to use this project to also introduce interrupts. We use the Atom commands for interrupts to instantly react to a Timer1 overflow thus making an automatic time base for flashing the LED.

The Atom runs on a 20 Mhz resonator clock signal. The external clock signal gets divided by four inside the Atom micro to form the internal clock pulse that drives the timers. We further divide that signal by 8 using the Timer1 prescaler. If you were to calculate this out you would see that the Timer1 will overflow every 0.104856 seconds.

$$\frac{65535}{20,000,000 \text{ Pulses/Second } (1/4) \ (1/8)} = 0.104856 \text{ seconds}$$

We want it to overflow on an even number such as 0.10 seconds or 100 milliseconds. If the Timer1 overflowed every 62500 pulses then it would be a perfect 100-msec time base. We can make this happen by presetting the Timer1 to 3035 (65535 – 62500 = 3035) or $0BDB hex. Then we track the number of overflows and when 10 have occurred we know that one second has passed so change the LED from off to on or on to off. That is what this project is all about.

$$\frac{62500}{20,000,000 \text{ Pulses/Second } (1/4) \ (1/8)} = 0.10 \text{ seconds}$$

Hardware setup

Hardware is exactly the same as Chapter 5 so this is an easy one to rebuild. If you saved the breadboard from Chapter 5 you can just move the Ultimate OEM module to that breadboard without having to wire up anything.

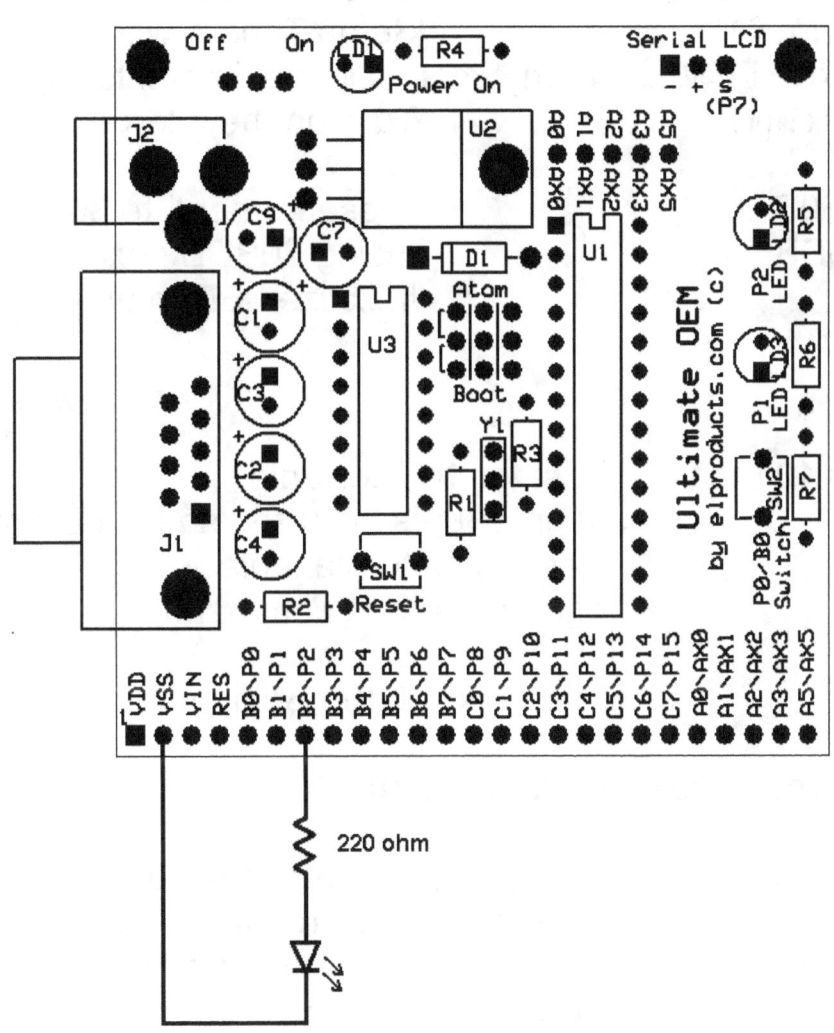

Software

The software is really not that long or really that complicated once you break it down. Remember, we are doing two advanced functions here, using the Timer1 and using interrupts.

```
counter var word
clear                              'Clear all variables
SETTMR1 TMR1INT8                   'Setup Timer1 with 1:8 prescaler
ONINTERRUPT TMR1INT,mytimer  ' Define timer interrupt
Enable TMR1INT                     'Turn on the interrupt

TMR1H = $0B                        'Preset Timer 1 to 3035
TMR1L = $DB                        ' using $0BDB hex
high p2                            'Initialize LED to on

main

if counter = 10 then       ' Test for 10 interrupts
toggle p2                  ' 10 interrupts occurred so flip LED state
counter = 0                ' Reset counter variable
endif                      ' end the If-Then command

goto main                  ' Loop back to the Beginning

'*** This is where we go on and interrupt ***

disable tmr1int            ' Prevent other interrupts from occuring
mytimer:                   ' Interrupt service routine label
TMR1H = $0B                ' Preset Timer 1 to 3035 decimal
TMR1L = $DB                '  using $0BDB hex
counter = counter +1       ' Increment the timer overflow count
resume                     ' This is how we exit an interrupt
```

How it works

The program starts off just like any other by establishing variables. We actually only use one variable and that is the byte variable "counter". Then we clear all the variables (in this case just one) with the CLEAR command.

```
counter var byte
clear                          'Clear all variables
```

Next we enter the special setup commands. As mentioned, the PIC automatically divides the 20 Mhz clock by 4 but we need to setup the divide by 8 prescaler. We do that with the SETTMR1 command and the TMR1INT8 option. This means, use the internal clock (20 Mhz divided by 4) and the 1:8 ratio prescaler as the Timer1 input.

```
SETTMR1 TMR1INT8               'Setup Timer1 with 1:8 prescaler
```

The interrupt has to be setup also and since there are various interrupt options we have to specify the Timer1 interrupt. We do this with the ONINTERRUPT command. The Timer1 interrupt name is TMR1INT so the command knows to interrupt on the Timer1 overflow. Finally the label of where to jump to when the interrupt or overflow occurs is defined as "mytimer". Later I'll explain what we do when the interrupt actually occurs.

```
ONINTERRUPT TMR1INT,mytimer  ' Define timer interrupt
```

Next, we turn the interrupt on with the ENABLE command. We have to specify which interrupt we want turned on so we add the TMR1INT selection.

```
Enable TMR1INT                 'Turn on the interrupt
```

The Timer1 is 16 bits long and is stored inside the PIC as two bytes. Therefore we preset Timer1 to 3035 decimal by directly changing the TMR1H register (high byte) and TMR1L register (low byte). The Atom has a RESETTMR1 command that will allow you to set both bytes at once to 3035 by issuing the command RESETTMR1 3035. This may be easier to use but I wanted you to understand that you can control the Timer registers directly. This makes you a better programmer in the long run.

```
TMR1H = $0B                    'Preset Timer 1 to 3035
TMR1L = $DB                    ' using $0BDB hex
```

We start the LED off in the on state by setting port P2 to a high value using the HIGH command.

```
high p2                        'Initialize LED to on
```

The main section of code starts with the label "main". This section is really simple. It checks if the variable "counter" is equal to 10 yet. If it isn't the program just loops back and does it again. If instead the value is equal to 10 then we want to change the state of the LED and we use the TOGGLE command to do that. TOGGLE just switches it from off to on or on to off.

Then we reset the "counter" variable to 0 and end the IF-Then statement with the ENDIF command. From there we loop back to "main" to test "counter" again.

```
main

if counter = 10 then    ' Test for 10 interrupts
toggle p2               ' 10 interrupts occurred so flip LED state
counter = 0             ' Reset counter variable
endif                   ' end the If-Then command
```

```
goto main                    ' Loop back to the Beginning
```

This last section of code is the interrupt code and is separate from the main loop of code. When the interrupt occurs, the program will finish whatever command was being executed and then jump to the defined interrupt label, which is "mytimer" in this example. Note that I said the Atom will finish its command before jumping to the interrupt label.

If you have a command such as PAUSE 5000 in your main loop, the interrupt will not occur until the full 5 seconds of pause has occurred. Therefore all commands should be short when you use interrupts. A FOR-NEXT loop of 5000 loops of PAUSE 1 would be a better way to achieve the same result but allow quicker interrupt response (also known as interrupt latency).

Before the "mytimer" label is the DISABLE command followed by the TMR1INT option. This shuts off the Timer1 interrupt for any code below that command. This is necessary so the interrupt cannot occur while we are running the interrupt routine. If we allowed that, we could end up in a continuous loop of interrupts and never leave the interrupt routine. This is also why it is very important to make your interrupt routines short.

'*** This is where we go on and interrupt ***

```
disable tmr1int              ' Prevent other interrupts from occurring
mytimer:                     ' Interrupt service routine label
```

Since the interrupt is shut off, you don't want your interrupt routine code to take so long that you miss an interrupt. Depending on what type of interrupt you are using, you will have to adjust the length of the interrupt service routine. Usually you can accomplish all you need in 10 short commands (no PAUSE 5000's) and never have any problem with any type of interrupt.

In this interrupt service routine, we do two things; reset Timer1 to 3035 and increment the "counter" variable. Remember we check this variable to see if it equals 10 in the main loop but we increment it in the interrupt service routine.

```
TMR1H = $0B              ' Preset Timer 1 to 3035 decimal
TMR1L = $DB              '  using $0BDB hex
counter = counter +1     ' Increment the timer overflow count
```

Finally we end the interrupt service routine with the RESUME command. All interrupt service routines must end with this command. It clears the interrupt flag inside the PIC and puts the program back at the command it was suppose to execute when the interrupt occurred.

```
resume                   ' This is how we exit an interrupt
```

Although this program isn't very long, it does demonstrate Timers and Interrupts quite well.

Next steps
You can easily change the IF-THEN test of "counter" to a larger or smaller value to make the LED flash quicker or slower. Another option is to change the preset values for TMR1H and TMR1L and prescaler to see if you can get it to flash the LED quicker or slower without changing the "counter" test value. If you put an oscilloscope probe on the LED you can see how accurate your calculations are.

You can take this same setup and modify it work with the Timer0, which overflows after 255 clock pulses. That will help you to prove you understand how this program works and at the same time develop a Timer0 sample program to use in the future. This is an advanced project so don't be too hard on yourself if it takes a while to understand.

Chapter 17 – External Interrupt on P0

Description

This final project will definitely be useful to the Atom user. The Atom offers all the PIC features as mentioned earlier via easy to implement commands. This project shows how easy it is to use the external interrupt (EXTINT interrupt) in Atom Basic. The external interrupt is at pin P0 that is internally connected to the PortB B0 PIC pin. This interrupt is probably the most useful and one of the easiest to use and understand. This project uses that interrupt in a way many beginners would not have thought of, as a multiplexed interrupt.

The hardware connections tie several inputs to the P0 pin thru diodes so different I/O pins can activate the external interrupt. In fact the project has four switches connected to the P4 thru P7 pins and all of them connected to the P0 pin thru diodes. When a switch is pressed the Atom program is interrupted from what it was doing and reads the P4 to P7 ports to see which switch was pressed. Then it lights the LED(s) that line up with the switch position(s) to show which switch or switches were pressed. While all this is going on, the Atom flashes a separate LED in the main loop to represent other functions that can happen while waiting for the interrupt to occur. The picture on the previous page shows the setup on the breadboard.

Hardware setup

The schematic shows the connections for this project. The four switches are tied to P4 thru P7 with a pull-up resistor to Vdd (5 volts). All the switches are connected to the P0 external interrupt pins thru the diodes. The diodes have the anodes tied to the B0 pin and the cathode connected to the switches. This allows the B0 pin to see a low (0.7v) signal when a switch is pressed. The LED's that indicate which switch was pressed are connected to the C4 thru C7 pins. The LED connected to the C0 pin is the continuously flashing LED in the project picture.

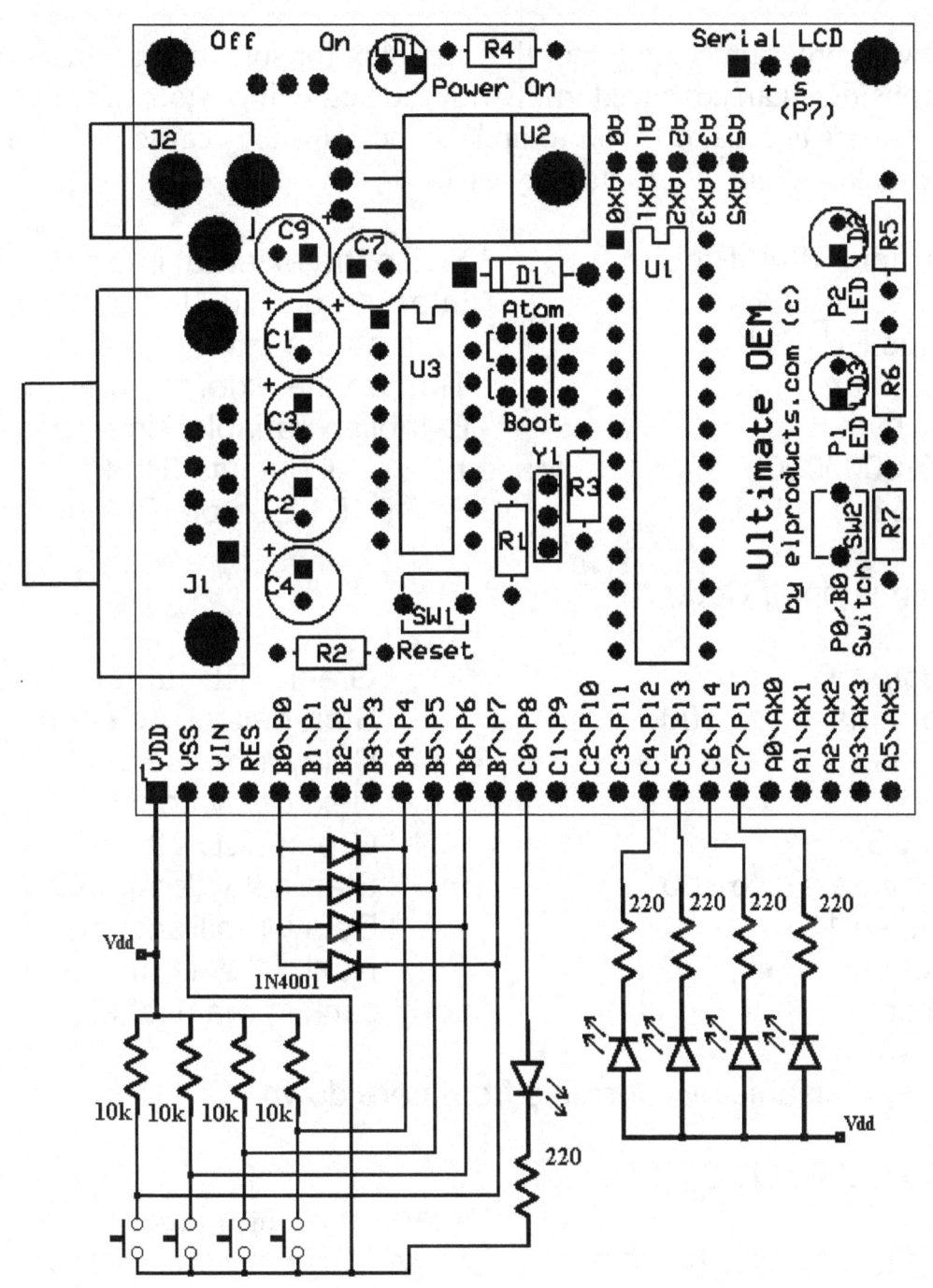

Software

The software program is really not that complex for something so useful. This is considered an advanced to PIC for the beginning Atom user but you will see it's not that difficult to understand. This is because the Atom software makes using interrupts very easy.

```
OnInterrupt ExtInt, ProgInt       ' Setup the external interrupt
setextint EXT_H2L                 ' Interrupt on High to Low signal
setpullups PU_ON                  ' Turn on the internal pull-ups
enable ExtInt                     ' Turn on the external interrupt
time var byte                     ' Establish variable Time
trisc = %00000000            ' Make port C all outputs (P8-P15)
portc = %11111111                 ' P12-P15 LEDs off, P8 on

'**** Main Loop of Code ******
Main
      high 8                      ' Green LED turned on
      for time = 1 to 100         ' Start delay loop count
      pause 1                     ' Delay 1 millisecond
      next                        ' Next delay count
      low 8                       ' Green LED off
      for time = 1 to 100         ' Start delay loop count
      pause 1                     ' Delay 1 millisecond
      next                        ' Next delay count
Goto Main                         ' Loop back to main label

disable       ' disable all interrupts from here down

'**** Interrupt Routine ******
ProgInt                           ' Interrupt routine label
portc.highnib = portb.highnib     ' Make LEDs match switches

hold                              ' label for below
```

```
if portb.highnib < %1111 then hold ' Wait for switches to be released
Resume                                ' This is how to exit interrupt
```

How it works

The program first establishes and sets up the external interrupt and defines the label of where to go when the external interrupt occurs.

```
OnInterrupt ExtInt, ProgInt          ' Setup the external interrupt
```

The external interrupt can happen when the P0 port transitions from a low to high state or a high to low state. We choose high to low with the SETEXTINT command and the EXT_H2L option. This will make the interrupt happen when we press the switch rather than when we let it go (EXT_L2H).

```
setextint EXT_H2L                    ' Interrupt on High to Low signal
```

Even though I show pull-up resistors on the schematic for the switches, I initially didn't include one for the P0 pin, which was a mistake. It needs one to make sure the P0 pin is sitting at a known state so I turned the internal pull-up resistors on in the software, which is available for P0 thru P7 only. I could have left off the switch pull-ups after that but they were there so I left them in. It works either way but this was a way of showing how to use the SETPULLUPS command and also add additional current drive from an external pull-up.

```
setpullups PU_ON                     ' Turn on the internal pull-ups
```

Now we turn the External Interrupt on with the ENABLE command and the EXTINT option.

```
enable ExtInt                        ' Turn on the external interrupt
```

A little different than the last project, I create the variable "time" after setting up the interrupt rather than before. I really don't have a great reason for this other than to prove the order doesn't really matter.

```
time var byte                          ' Establish variable Time
```

Another different thing I did is write to the port registers directly (TRISC and PORTC) to setup the P8, P12 thru P15 ports. This is just an easy way to do it and is the same way you would do this if you were programming the PIC directly in another language. Just a fine example of how the Atom gives you full control of the Microchip PIC 16F876.

```
trisc = %00000000          ' Make port C all outputs (P8-P15)
portc = %11111111          ' P12-P15 LEDs off, P8 on
```

The main loop of code once again starts with the "main" label. You can call it what you want but this makes it easier for me to understand when I look at my code many months later. The main loop just flashes the green LED on port P8 on and off at a 100-millisecond rate. The reason it looks so long is because I break up the 100-millisecond delay into a FOR-NEXT loop with a 1-millisecond delay repeated 100 times. I do this because of the interrupt.

```
Main
        high 8                          ' Green LED turned on
        for time = 1 to 100             ' Start delay loop count
        pause 1                         ' Delay 1 millisecond
        next                            ' Next delay count
        low 8                           ' Green LED off
        for time = 1 to 100             ' Start delay loop count
        pause 1                         ' Delay 1 millisecond
        next                            ' Next delay count
Goto Main                               ' Loop back to main label
```

When an interrupt occurs, the Atom will finish the command it is working on before jumping to the interrupt service routine. If I just used PAUSE 100 as the 100-millisecond delay then the interrupt could occur when the command started and the interrupt routine would not get processed until 100-milliseconds later. By then the switch could have be released and the software would not know which switch was pressed. This is why the delay is broken into several commands that take very little time to execute.

The interrupt routine is very short but I do something not normally done in an interrupt routine, I force it to stay there until the switch is released. First the DISABLE command is issued to indicate all the commands below it cannot be interrupted.

```
disable        ' disable all interrupts from here down
```

The interrupt service routine starts at label "ProgInt". The first command is complex looking but very simple. I use the Atom option of reading and writing to the registers directly. I read the switches P4 thru P7, which form the bits of register PortB's high nibble (upper 4 bits). Using the equal sign I make P12 thru P15, which form the PortC high nibble, the same state as the bits of PortB. Thus, I'm making the LEDs match the state of the switches. Those that are pressed (low) light the LEDs on P12 thru P15. Those that are not pressed (high) turn off those LEDs on P12 thru P15. And I did all that in one command. Easy huh?

```
ProgInt                          ' Interrupt routine label
portc.highnib = portb.highnib    ' Make LEDs match switches
```

Then I test the PortB high nibble to see if any of the bits are "0" indicating a switch is being held closed. I do this in a continuous loop by jumping back to the hold label until all the switches are released. When the switches are all released, the RESUME command is executed and the program jumps back to the main loop where it was interrupted.

```
hold                              ' label for below
if portb.highnib < %1111 then hold ' Wait for switches to be released

Resume                            ' This is how to exit interrupt
```

Next steps
One thing you can try is to replace the main loop of code with something more interesting than flashing an LED. For example you could add software to drive an LCD display that shows how many times the switch was pressed. That can be expanded on to do all kinds of things.

The switches can be replaced with light sensors, connected to A/D pins, that sense when someone walks by. This way you are creating a people counter with the total count shown on the LCD screen. You see how all these projects come together?

Chapter 18 – Conclusion

I never really know how to end a book because I have so many things I could do but I run out of time and pages. Hopefully you now feel comfortable attacking just about any project with the Atom. I hope the projects and explanations I've covered helped you learn to program the Atom modules better than just reading the manual. I still highly recommend you read the manual but having actual projects to do is a lot more fun to me.

From here you should start thinking about those great inventions you can create around the Atom micro. An 18 pin chip is in the works as I write this so hopefully you will have three choices soon; an 18 pin, 28 pin and 40 pin version to design with.

If you have any questions or comments (good or bad) send them to me via email at chuck@elproducts.com. It's the feedback from readers that allows me to know if my work was successful or not. Book sales only tell you that you people were interested enough to buy the book. It's the feedback that tells you if you actually did something good (or bad). I've gotten a lot of emails from my first book that told me I helped people learn to become programmers and also got them back into the electronics industry. That is priceless feedback. Send me your story.

Finally, thank you for buying and/or reading this book. If I get enough positive responses from it, I hope to do another book of just Atom projects. I've got so many projects that just would not fit in here. Email me what you would like to see. Maybe I will put your idea in the next book. So long for now and have fun.

Appendix A – Sources

Atom, Ultimate OEM and BasicBoard and other Microchip PIC Products

Beginner Electronics
www.beginnerelectronics.com
beginnerelectronics@gmail.com

Basic Micro
www.basicmicro.com

Light Sensor Module

www.phidgets.com

Max7219 LED Driver

www.maxim-ic.com

LCD, Misc Components

Beginner Electronics
www.beginnerelectronics.com
beginnerelectronics@gmail.com

www.jameco.com

Appendix B – Referenced Files

The files reference throughout the book along with the project software can be downloaded from my website at:

www.elproducts.com/atombookfiles

The files are combined in a .zip file and include:

- **Atom Compiler Software**
- **Atom Manual**
- **Ultimate OEM Manual**
- **PIC16F87X(A) Data Sheet**
- **PIC16F88x Data Sheet**
- **PIC16F88 Data Sheet**
- **Max7219 Data Sheet**
- **44780 LCD Driver Data Sheet**
- **Project Software**

Index

E

F

G

H

I

L